Can Kissing Make You Live Longer?

Can Kissing Make You Live Longer?

BODY AND BEHAVIOUR MYSTERIES EXPLAINED

Dr Stephen Juan

HarperCollins*Publishers*

HarperCollins_Publishers_

First published in Australia in 2010
by HarperCollins_Publishers_ Australia Pty Limited
ABN 36 009 913 517
harpercollins.com.au

HarperCollins_Publishers_
25 Ryde Road, Pymble, Sydney, NSW 2073, Australia
31 View Road, Glenfield, Auckland 0627, New Zealand
A 53, Sector 57, Noida, UP, India
77–85 Fulham Palace Road, London, W6 8JB, United Kingdom
2 Bloor Street East, 20th floor, Toronto, Ontario M4W 1A8, Canada
10 East 53rd Street, New York NY 10022, USA

National Library of Australia Cataloguing-in-Publication data:

Juan, Stephen
 Can kissing make you live longer? : body and behaviour mysteries / Dr Stephen Juan.
 ISBN: 978 0 7322 9050 4 (pbk.)
 Includes index. Bibliography.
 Body, Human – Miscellanea.
612.002

Cover design by Jane Waterhouse, HarperCollins Design Studio
Cover images by shutterstock.com
Internal design by Alicia Freile, Tango Media
Typeset in 11/16pt Dante MT Regular by Kirby Jones

To the many readers who have asked questions

Contents

Introduction   1

Chapter 1   Beginnings   3
Chapter 2   The Head   28
Chapter 3   The Eyes   44
Chapter 4   The Nose   58
Chapter 5   The Ears   72
Chapter 6   The Mouth   82
Chapter 7   The Skin   103
Chapter 8   The Hair & Nails   127
Chapter 9   The Skeleton, Bones & Teeth   136
Chapter 10   The Heart, Blood & Lungs   150
Chapter 11   The Stomach & Intestines   167
Chapter 12   Otherwise Inside   183
Chapter 13   Behaviour   204
Chapter 14   Endings   240

Afterword   263
References   264
Acknowledgements   294
About the Author   297
Index   298

Introduction

Greetings, reader!

Welcome to my latest book, packed full of lots of curious and strange facts and information about our fascinating human body. No doubt there will be something to interest everybody, or should I say every *body*?

For starters, find out why fewer boy babies are born during hard times, why 'morning sickness' is necessary during pregnancy, and the difference between a chromosome and a gene (Chapter 1). Find out what a 'brain freeze' is and why it hurts so much (Chapter 2), how far the naked eye can see (Chapter 3), and why pepper makes you sneeze (Chapter 4). Discover if our ears grow longer with age (Chapter 5), if it is possible to get a food allergy by kissing someone (Chapter 6), and what manner and number of microscopic life live on the surface of the human body (Chapter 7). Learn why we are actually *not* 'naked' while in our mother's womb (Chapter 8), why some people cannot stop tapping their toes (Chapter 9), and what shape the human heart *really* most resembles (Chapter 10). Look into why opera singers are usually fat (Chapter 11), whether the human body generates light (Chapter 12), and if there is a limit to the number of times you can almost die and be brought back to life (Chapter 14).

And if all this doesn't arouse your interest, then there's a special chapter on human behaviour (Chapter 13). Of course it is purely coincidental that the behaviour chapter is number 13 and has nothing to do with anyone suffering from triskaidekaphobia (an irrational fear of the number 13). Chapter 13 explores topics such as why so many people believe in conspiracy theories, why so many gamble, and why some people are more accident prone than others. It's also quite curious to know why some people confess to crimes they did not commit, if lie detector tests really work, and why humans so often doodle, and much, much more.

There are probably many other intriguing topics not yet discussed, but we will hopefully get around to all of them eventually — quite a task! In the meantime, enjoy this latest offering.

Chapter 1
Beginnings

WHAT MAKES A HOMO SAPIEN A HOMO SAPIEN?
(Asked by Ralph Turner of Bangor, Maine, USA)

All forms of life, from the simplest to the most complicated, are classified scientifically into a system known as taxonomy. As humans we are *Homo sapiens* — we are animal, mammal and primate in one. According to the classic taxonomic system developed by Swedish biologist Carl Linnaeus (1707–1778), and added to by many others since, humans are classified into: domain, kingdom, phylum, class, order, superfamily, family, genus and species. Although some authorities dispute various aspects of what appears below and have even proposed additional categories, humans are taxonomically humans due to the following:

- *Domain: Eukaryota*. We are classified among organisms with a complex cell or cells and with genetic material that is organised into a membrane-enclosed nucleus or nuclei.
- *Kingdom: Animalia*. We are classified among organisms that are multi-cellular, capable of locomotion and feeding themselves via other organisms or parts of other organisms, and those that develop a stable body structure.
- *Phylum: Chordata*. We are classified among the vertebrates (those animals with a type of spinal cord) or closely related and complicated invertebrates. Chordates have a *notochord* which is a flexible, rod-shaped body form in embryos. In lower vertebrates it continues throughout life. In higher vertebrates it becomes a vertebral column. Chordates have a *dorsal nerve cord* which later

3

becomes the brain and spinal cord. It is formed from a part of the ectoderm that rolls and forms a hollow tube. Chordates have *pharyngeal slits* used for feeding. In primitive chordates the pharyngeal slits strain water and filter food particles. In more advanced chordates the pharyngeal slits disappear at the embryo stage. Chordates have an *endostyle* which is a groove on the ventral wall of the pharynx that produces mucus used in eating. Chordates have *pharyngeal pouches* which are also used for feeding. These become gills in fish. Chordates also have a tail. In our case we have only vestigial evidence of this.

● *Class: Mammalia.* We are classified among the vertebrate animals that have: mammary glands that produce milk; hair or fur; specialised teeth; three small bones in the ear; a neocortex region in the brain; 'warm-blooded' bodies; a four-chambered heart; and a brain-regulated circulation and temperature-controlling system.

● *Order: Primate.* From the Latin word *primus* meaning 'first', we are classified among the *prosimians* that most closely resemble the early proto-primates (such as lemurs), the 'monkeys of the New World' (such as capuchin monkeys), and the 'monkeys of the Old World' (such as baboons).

● *Superfamily: Hominoid.* We are classified among the non-human-like primates, that is, the great apes — chimpanzees, gorillas and orang-utans.

● *Family: Hominidae.* We are classified among the human-like primates both extinct and still in existence.

● *Genus: Homo.* With the Latin *homo* meaning 'man' or 'human' in its more recent meaning, we are classified among modern humans and our close now extinct relatives. *Homo erectus* (upright human), *Homo neanderthalis* (Neanderthal human), *Homo habilis* (tool-making human), *Homo floresiensis* (flower human) and others.

● *Species: Sapiens.* With the Latin *sapiens* meaning 'wise', we are *Homo sapiens* (wise humans). Some argue that *Homo sapiens idaltu* (elder wise humans) is of our species, too, although also now extinct; others argue that we should be rightly classified as *Homo sapiens sapiens* to make clear this last distinction.

WHAT IS THE NEXT STEP IN HUMAN EVOLUTION FOR *HOMO SAPIENS*?
(Asked by Ralph Turner of Bangor, Maine, USA)

It is anyone's guess as to what will be the next step in the evolution of *Homo sapiens*. Nature has been working on the model for more than 3 million years through natural selection. Some scientists have speculated that the next adaptive changes to humans may be to the anatomical structure with better spine, ears, eyes, hips, joints, etc. or perhaps it will be changes in our brain or in our consciousness that will usher in the next subspecies. *Homo sapiens humanus* (as opposed to *Homo sapiens non-humanus*) would be a modern human who is humane, cultured, refined, educated, compassionate, empathetic, creative, intuitive, courageous, spiritual, generous and energetic. (An ideal dinner party guest?) They would be teachers, healers, helpers, creators, preparers, nurturers, transformers and repairers. (Pretty handy to have around, right?) They would see something and want to make it better, and think of others not just of themselves. They would understand the most important philosophical principle of human awareness: that the greatest receiving comes from the greatest giving. Our body, brain and consciousness would evolve together. For example, we now know that experience during the early years of human development has effects on brain structure, particularly in the frontal lobe region responsible for empathy. This in turn affects changes in human behaviour. Those receiving a *humanus* experience will develop more advanced brains than those receiving a *non-humanus* experience. Such advancement would be passed on from offspring down to another and so on in the lineage. Perhaps through our more enlightened behaviour we could influence our own evolution?

HOW WAS PATERNITY DETERMINED BEFORE WE HAD DNA ANALYSIS?
(Asked by Shawn Joseph of Rockdale, New South Wales)

Before DNA analysis, paternity was established by comparing inherited characteristics of blood cells from the father, mother and child. For example, the membranes of white blood cells contain surface proteins known as antigens. People who are not related almost never have the same type of antigen. So the most accurate paternity tests used antigens on white blood cells called leucocytes. On average, using this test ruled out 90 per cent or more of men as a child's father. Other tests could be performed on the antigens and internal proteins of the red blood cells and the proteins in plasma, the clear part of the blood. A thorough paternity test consisted of some 34 different tests that together were about 99.6 per cent accurate. If a man was not excluded through these tests, the probability of paternity was calculated by comparing the likelihood that the child inherited shared characteristics from him rather than from an unrelated, unknown man of the same ethnic background. If the probability was above 90 per cent, then most courts held that the man was the child's father.[1]

WHEN DID HUMANS FIRST START GIVING BIRTH BY CAESAREAN?
(Asked by Rene Bernard of Vancouver, British Columbia, Canada)

Ancient Egyptian folklore contains references to Caesarean section (CS) deliveries going back about 5000 years. Throughout most of history, CS was performed only as a last resort and to save the baby from a dying mother. The mother *did* almost always die. CS was so dangerous that the mother probably considered herself lucky if she and her baby survived, even if she was rendered infertile, which also often happened.

The popular belief exists that the Roman emperor Julius Caesar was born by CS and gave his name to it. But this is unlikely. Aurelia, Caesar's mother, survived. This is unusual in itself. Also, under Roman law at the time, CS would not have been performed on her. Of course those attending her may have decided to ignore the law to save the mother and baby. Later, the Roman law governing CS known as *lex regia* was changed to *lex caesarea* under subsequent Roman emperors. This is probably how the name came about.

It was not until the 16th century that the first case of a woman giving birth by CS and surviving was verified. However, if Aurelia did give birth by CS, it is even more remarkable since she gave birth to further children. Ironically, under Roman law, the mother's husband performed the surgery. In that sense Caesar would have been fortunate. His father was a swineherd. Swineherds often have experience performing similar surgery on pigs. In England and North America, death rates for the mother in CS deliveries were about 75 per cent until the mid-19th century. For the baby, it was only a little better.[2]

WHY ARE THERE MORE CAESAREAN BIRTHS NOW?
(Asked by Rene Bernard of Vancouver, British Columbia, Canada)

Although CS is major abdominal surgery, it is generally regarded as being safer now than at any other time in our history. This is due to overcoming post-operative complications including the ability to fight infections. Prolonged labour and risks to both mother and baby places pressure on childbirth professionals to act quickly. Statistically, in the English-speaking world, women are waiting until later in life to have children. As a consequence, more women experience natural childbirth difficulties due to causes not entirely understood. CS deliveries far exceed the estimated 12 per cent incidence of need. Yet in some areas of the English-speaking world, the CS figure is 30 per cent or more. Beyond this, in Brazil CS delivery is regarded as the *preferred* form of delivery and is taught so in medical schools. A team of US doctors in California argues that the recent assertion of 'a woman's right to choose' is a major factor in 'CS or no CS'.[3]

◆

The most foetuses found in a human body was 15. This included 10 girls and 5 boys. They were 4 months old when they were removed from the womb of an Italian housewife in July 1971. The woman had been taking a fertility drug.

◆

Thanks to the twisting and cramped interior of the birth canal, human babies, unlike those of other primates, tend to turn mid-birth and exit the vagina facing downwards.

◆

The World Health Organization estimates that worldwide 15 per cent of childbirth labours have a life-threatening complication.

◆

It has been estimated that the 'natural' rate of maternal death from childbirth is between 1 and 1.5 per cent. The biggest risk is uncontrolled bleeding.

HAS MUTILATING THE BODIES OF CHILDREN EXISTED IN ALL SOCIETIES THROUGHOUT HISTORY?
(Asked by Miguel Padilla of Mexico City, Mexico)

If not all, then certainly most societies have routinely mutilated the bodies of children. Of course this depends on what one considers as 'mutilation'. For example, the practice of circumcision is thought of as mutilation by some people yet not by others. If we define it broadly, then mutilation of children's bodies in some form or another is found in nearly all cultures throughout history. For instance, the prehistoric middle Palaeolithic period began some 300,000 years ago. *Palaeo* is Greek for 'old' and *lithos* is Greek for 'stone', hence 'stone age'. Palaeolithic cave paintings show children's hand prints often with fingers missing. It is believed that children had their fingers cut off in the widespread belief by many cultures that animal spirits demanded a child's finger to appease them. Finger sacrifice rituals have been found in many cultures as have places where amputated fingers were stored.[4, 5]

HOW DO WE DEFINE 'RACE'?
(Asked by Sean Scully of Riverview, New Brunswick, Canada)

A way of thinking about 'race' is to define it as including all members of any human group that can successfully breed with each other. A 'race' includes all the descendants of a common human ancestor. So it is like a family, a tribe, an ethnic group, a nation, or the entire world's human population. This is the broadest biologically defensible definition. 'Race', as it is commonly understood (and misunderstood) is a social construct rather than a biological one. It is impossible to scientifically define as distinct a human race as apart from another distinct human race on biological criteria alone. There are as many human races as there are human beings due to the genetic uniqueness of each of us. The only possible exception to this would be identical twins, where the number of members of that race would be exactly two!

There are genetic-based outcomes among populations living in specific geographical areas that can set them off from populations in other geographical areas. These outcomes may make certain populations more or less susceptible to having specific anatomical features or health conditions. There may be a survival advantage favouring a particular feature in a certain environment. For instance, more prominent eyelid folds among the Inupiak people of the Arctic may help protect them against snow blindness. Such an anatomical feature would be of no advantage in a snow-free environment. Tay-Sachs disease, a genetic-based disease causing mental and physical deterioration and often death by age 4, is more prominent among Eastern European Jews. Sickle Cell Anaemia, a disease affecting red blood cells that results in early death (often by age 40), is more prominent among sub-Saharan Africans.

The rate of giving birth to twins varies with geography. West Africa has the highest birth rate of twins in the world. For example, the Igbo-Ora of Nigeria have 31.6 twin births per 200 live births compared to an average of 1 twin birth per 200 live births in European nations.

The fact that all anatomical features or health conditions differ among human populations proves only that there is variation everywhere on Earth. But to define any one population as a 'race' as apart from another population as another 'race' is to divide people on a crude and misleading basis. Such a judgment invites bias and prejudice.

'Race' has been given great prominence in Western history, particularly over the last 600 years with the advent of New World exploration, the contact of Europeans with non-Europeans, and European empire building. 'Race' has often been the ill-conceived basis for the most terrible discriminatory policies, laws, social movements, ethnic cleansings and wars in the long and sorry history of man's inhumanity to man. 'Race' is today a scientifically and socially bankrupt concept.[6]

◆

There is a 1 in 200 chance that a male human worldwide is a direct patrilineal descendant of Genghis Khan.

◆

The man with the oldest known individual ancestor is British professor Adrian Targett. DNA tests matched Professor Targett's with that of a 9000-year-old skeleton found in Cheddar, England.

◆

The oldest authenticated pair of female twins was Kim Narita and Gin Kanie of Japan. They were born on 1 August 1892. Kim died of heart failure in January 2000 at 107 years of age.

◆

The longest living triplets were Faith, Hope and Charity, who were born in Elm Mott, Texas, on 18 May 1899. Faith was the first to die at age 95 in October 1994.

◆

The world's oldest quadruplets were the Ottman siblings — Adolf, Anne-Marie, Emma and Elisabeth. They were born on 5 May 1912. Adolf died first at age 79 in March 1992.

CAN AN EXTINCT HUMAN LIKE A NEANDERTHAL MAN BE CLONED AND BROUGHT BACK TO LIFE?
(Asked by Ken Alvarez of Shreveport, Louisiana, USA)

Theoretically, this is certainly possible, but there would be technical problems along the way in trying to do it. There are also many ethical and moral issues to be resolved satisfactorily beforehand. After all of this, the steps for cloning a Neanderthal would be similar to the following:

1. Obtain a reliable DNA (deoxyribonucleic acid) sample. A thorough search is required for cells from the to-be-cloned Neanderthal, including bone and teeth cells. Neanderthal hair and skin cells could be used if available, but since hair and skin make poor fossils compared with bones and teeth, none survive for the Neanderthal. Extracting healthy DNA fragments from the nucleus of cells of Neanderthal bones and teeth would be very difficult to do.
2. Rebuild the genome. The broken DNA of the Neanderthal is reassembled using the genome of a related living human as a guide.
3. Swapping DNA. Eggs from the ovaries of a living human are removed and their nuclei replaced with the restored genetic material from the Neanderthal.
4. Stimulating the eggs. Genetically engineered restructured eggs are treated with chemicals or perhaps electric current to fuse the nuclei with the eggs and trigger cell division.

5. Implanting the embryo. After the desired cell division has started and progressed to form an embryo of about 200 cells, the embryo is implanted into the womb of a living woman where it is carried to full term like any other human embryo.

6. Birth. The surrogate mother gives birth to the Neanderthal baby in a normal human birth, and the baby will grow up into a Neanderthal adult.[7, 8, ,9]

WHAT IS THE DIFFERENCE BETWEEN A GENE AND A CHROMOSOME?
(Asked by Lynn Davis of Casper, Wyoming, USA)

All life is made up of cells. All cells in the human body, except red blood cells, contain chromosomes. Chromosome comes from the Greek *khroma* meaning 'colour' and *soma* meaning 'body'. Chromosomes got their name from the first lab experiments in the 1880s where it was revealed that chromosomes could be stained with dyes, making them easier to study.

A gene is located on a chromosome. Every factor in inheritance is due to a particular gene. Genes specify the structure of particular proteins that make up each cell. Gene comes from the Greek *genea* meaning 'generation', 'origin', 'beginning', 'kin' or sometimes 'race'. Gene was shortened from 'pangene' which means 'all-generation'. Genes contain DNA, the chemical basis of heredity.

Think of it this way: DNA is in genes, genes are on chromosomes. When it was first seriously conceived to 'map' all genes on all human chromosomes, it was called the Human Genome Project — a combination of 'gene' and 'chromosome'.[10]

WHAT HAPPENS WHEN A CHROMOSOME IS ABNORMAL?
(Asked by Lynn Davis of Casper, Wyoming, USA)

Each chromosome has a specific and proper structure and colour. Together, all the chromosomes in our body form a distinct pattern called the human karyotype. We have 46 chromosomes (23 from each parent), however there are variations in reproduction due to abnormalities in the karyotype of an individual. A few of these abnormalities are beneficial, most are harmless, and a few are catastrophic. A major chromosome abnormality (MCA) accounts for half of all spontaneous human abortions. An MCA occurs in about 1 in every 100–200 births. Recent medical diagnostic techniques can now detect many MCAs long before the child is born by analysing the patient's blood under a microscope. MCAs take one of five forms:

1. In **duplication**, instead of the normal two, the individual has three copies of the chromosome in every cell of their body (trisomic). Most trisomics result in a spontaneous abortion — but not always. Trisomy-21 (Down's syndrome) occurs 1.5 times in every 1000 births — the individual has three copies of chromosome 21. Trisomy-8 (Edward's syndrome) occurs 3 times in every 10,000 births — the individual has three copies of chromosome 18. Trisomy-13 (Patau's syndrome) occurs 2 times in every 10,000 births — the individual has three copies of chromosome 13.

2. In **deletion**, a piece of chromosome may go missing. The missing piece is usually from the end of the chromosome, but can be missing from the middle as well. The types of symptoms and their severity will vary depending on the size and location of the deletion. Deletion syndromes occur at a frequency of about 1 in every 16,000 births. In Wolf-Hirschhorn syndrome (1 in every 50,000 births) the abnormality occurs on chromosome 4. In Cri du Chat syndrome (1 in every 50,000 births), the abnormality occurs on chromosome 5.

3. In **translocation**, a whole or a piece of a chromosome becomes attached to another.

4. In **inversion**, a portion of the chromosome breaks off, turns upside down, and reattaches.

5. In **rings**, a portion of the chromosome breaks off and forms a circle, with or without loss of genetic information.

In the last three forms, there is an alteration in the pattern of genetic information.[11]

CAN BRAIN DAMAGE OCCUR IF A BABY IS LEFT TO 'CRY IT OUT'?
(Asked by Christine Koch of Strathfield, New South Wales)

Research suggests that allowing a baby to 'cry it out' can cause brain damage and *at best* causes extreme distress to the baby. Evidence is mounting that such distress in a newborn blocks the full development of certain areas of the brain and causes the brain to produce extra amounts of cortisol which can be harmful. According to Dr Michael De Bellis,[12] children who suffer early trauma generally develop smaller brains.[13]

Dr Martin Teicher[14] and colleagues claim that the brain areas affected by severe distress are the limbic system, the left hemisphere and the corpus callosum. They note that baby neglect resulted in a 15 to 18 per cent reduction in regions 3, 5 and 7 of the corpus callosum (the connective fissure between the two hemispheres of the brain that facilitates their communication).[15] Additional areas that may be involved are the hippocampus (located in the centre of the brain and vital for long-term memory and spatial perception) and the orbitofrontal cortex (located at the front of the brain and vital for thinking such as decision making).

Dr Margot Sunderland[16] argues that some of the brain-damaging effects may occur if parents fail to properly nurture a baby — and that means *not* allowing them to 'cry it out'. Dr Sunderland draws upon her work in neuroscience to come to these conclusions and recommendations about parenting practice. In *The Science of Parenting*, the first parenting book to link parent behaviour with infant brain development, Dr Sunderland describes how the infant's brain is still being 'sculpted' after birth, and that parents have a major role in the brain 'sculpting' process. To do this properly, Dr Sunderland advocates that it is crucial for parents to meet the reasonable emotional needs of their child. This is assisted by continuously providing an emotionally nurturing environment for the child. Allowing a baby to 'cry it out' when they are upset will probably be regarded as child abuse by future generations.[17, 18]

WHY IS IT SO HARD TO LET A CRYING BABY CRY?
(Asked by Nikki Ryan of Toowoomba, Queensland)

Research shows that our 'cry response' to a child's cry may be hard wired into our brain. A baby cries because it is hungry, tired, too hot, too cold, lonely or otherwise unhappy. In effect it is the only way the child can communicate their needs. If they could talk, they would. If a baby did not have a terribly shrill cry, they would less likely get attention and, from an evolutionary point of view, would unlikely be responded to and hence perish. So biologically, nature gives a baby their terrible cry for survival purposes. According to Anni Gethin and Beth Macgregor,[19] 'Parents are too biologically wired to not respond to their babies' cries; responding isn't simply a matter of choice, like whether to go see a movie or a band on a Friday night. Magnetic resonance images (MRIs) of mothers' brains show that the hypothalamus and cingulated (parts of the mammal brain) are activated when their baby cries. The result is that most mothers feel physically compelled to pick up and soothe their crying babies ...'[20]

DO HAPPIER BABIES GROW UP TO BE BETTER LOVERS?
(Asked by Nikki Ryan of Toowoomba, Queensland)

A recent study suggests that happier babies make better romantic partners later in life. Dr Jeffry Simpson[21] and his team studied a group of 78 babies from age 1 over a period of 25 years. In a series of experiments on 'attachment', the Simpson team first noted which babies were more 'securely attached' to their mothers (and hence 'happier') compared with other babies. Follow-up studies many years later when the babies grew up to around age 20 to 23 years old showed (all other things being equal) that the adults who were more secure as babies had much deeper and overall happier romantic relationships than those who were less secure as babies.

The Simpson team adds that the babies who were more securely attached to their mothers had 'more secure relationships with close friends at age 16, which 'in turn predicted more positive daily emotional experiences in their adult romantic relationships (both self- and partner-reported) and less negative effect in conflict resolution and collaborative tasks with their romantic partners (rated by observers).'[22]

So how do you get 'more securely attached babies'? For starters, always pick them up when they cry.

IS AMNIOTIC FLUID SIMPLY URINE?
(Asked by Giulia Rossi of Rome, Italy)

Amniotic fluid is the clear, yellow-brownish liquid that surrounds the unborn baby; it is contained in the amniotic sac. The fluid consists mostly of foetal urine that bathes the developing foetus. The foetus floats in this fluid — warmed to the mother's body temperature. The amniotic fluid increases in volume as the foetus grows, with the greatest amount at about 34 weeks GA (gestational age). At 34 weeks GA, the amount of amniotic fluid is about 800 ml. This amount reduces to about 600 ml at 40 weeks GA when the baby is born. Throughout the mother's pregnancy amniotic fluid is continually being swallowed and 'inhaled' and replaced through being 'exhaled', as well as being urinated by the baby. Though the fluid would not be terribly tasteful to us, the developing baby does not seem to mind it at all. It is essential that the amniotic fluid be breathed into the lungs by the foetus in order for the lungs to develop normally.

Analysis of amniotic fluid, drawn out of the mother's abdomen in an amniocentesis procedure, can reveal many aspects of the baby's genetic health. This is because the fluid also contains foetal cells which can be examined for genetic defects. Amniotic fluid also protects the developing baby by cushioning it against blows to the mother's abdomen, allows for easier foetal movement, promotes muscular or skeletal development and helps protect the foetus from heat loss. Amniotic fluid? We all knew it very well — it used to be all around us.[23, 24]

◆

For roughly 6 to 7 months after birth, an infant can breathe and swallow at the same time. This ability was beneficial before the infant was born when it lived in the watery environment of the mother's womb.[25, 26]

◆

The navel divides a baby directly in half by length.

WHY CAN'T I REMEMBER BEING BORN?
(Asked by Jon Bovard of London, UK)

The subject of birth memory is very controversial among human development scientists. Some authorities argue that people *can* remember their own birth, the first year after birth, and even pre-birth memories. This is accomplished usually by methods such as 're-birthing', use of a 'primal' therapist, a 'dianetics auditor', hypnosis, dream analysis and deep meditation. Boris Brott, the Canadian conductor and motivational speaker, claims he discovered he could play certain pieces sight unseen due to the fact that his mother, a professional cellist, had practised these same pieces over and over during her pregnancy. In conducting a score for the first time, the cello parts would sometimes 'jump out at him'. He knew how they went before turning the page of music. Yet other authorities, who dismiss claims such as Brott's, argue that people cannot remember their own birth since the human brain is too underdeveloped for memory so early. They also contend that such sensory data, even if taken in, are quickly lost as memories because the foetus and newborn have no words to 'hang onto' them. However, there are problems explaining lost memories of sounds, smells, tastes and other sensations that may be less reliant upon words.

Of the many fascinating studies on the topic of foetus memory, one such study provides evidence that a baby can indeed possess some memory that involves words before birth. Psychologists Anthony DeCasper along with Melanie Spence[27] conducted a very simple experiment. They asked a group of pregnant women to read aloud *The Cat in the Hat* by Dr Seuss twice a day during the last 6 weeks of their pregnancy. A few days after birth, the babies were given the opportunity to hear recordings of two stories. One was the familiar Dr Seuss story. The other was another Dr Seuss story they had never heard before. Outfitted with earphones and a special nipple that allowed them to switch the story being heard by sucking faster or slower, 10 out of 12 newborns changed their speed of sucking to arrive at the familiar story, thereby rejecting the new story. This suggests that the babies could hear, differentiate between and remember stories. They also preferred the familiar story to the unfamiliar one. In essence, they voted 'with their mouths'.[28, 29, 30]

WHAT IS THE PURPOSE OF MORNING SICKNESS?
(Asked by Matt Gray of Canberra, ACT)

Morning sickness is recurrent nausea and vomiting often seen in the first 4 to 12 weeks of pregnancy. Approximately 60 to 80 per cent of pregnant women will experience it, and for about 20 per cent of these women the symptoms may extend beyond 12 weeks. Morning sickness is more pronounced in the morning than in the afternoon, evening or night time, hence its name. However, it can occur at any time of the day.

In their review article on morning sickness, Drs K.Y. Loh and N. Sivalingam[31] write that 'nausea and vomiting are usually mild and self-limiting, however some of the mothers have a more profound course which lead to *hyperemesis gravidarum*'.[32] *Hyperemesis gravidarum* (HG) is the most severe form of morning sickness. About 1 in every 1000 pregnant women suffer from HG. *Hyper* means 'excessive', *emesis* means 'vomiting', *gravida* means 'pregnant woman', and *rum* in this sense means 'in danger'; hence an 'excessively vomiting pregnant woman in danger' — a very apt description. HG involves repeated vomiting, dehydration and weight loss. Hospitalisation and the administration of intravenous fluids and monitored feeding may be required in extreme cases. Drs J.D. Quinla and D.A. Hill[33] write that 'although several theories have been proposed, the exact cause [of morning sickness] remains unclear'.[34] Among the proposed theories, morning sickness symptoms could be triggered by:

- high levels of hormones, including oestrogen
- fluctuations in blood pressure, especially low blood pressure
- *Helicobacter pylori*
- altered metabolism of carbohydrates
- interplay of physical and chemical changes
- protection from toxic substances.

With respect to this last theory, it is possible that a pregnant woman may have developed an aversion to foods more likely to harm the developing foetus. It is in the first 12 weeks that damage can be done most easily to the baby's developing central nervous system. So if the mother keeps her diet very simple with 'safe' foods, the baby is protected. As Drs Gillian Pepper and S. Craig Roberts[35] point out morning sickness 'serves an adaptive prophylactic function against potentially harmful foodstuffs'.[36]

WHY DOES NATURAL SELECTION TAKE SO LONG TO GET RESULTS?
(Asked by Colin Jackson of Telford, Shropshire, UK)

Colin further asks, '... when controlled breeding programs can get results in a relatively much shorter period of time, wouldn't more rapid evolution itself be a good species survival trait and so have been selected for?'

Although there is fierce debate about how fast natural selection can proceed and if natural selection is still proceeding in humans due to our technology, there is a short answer. Speed is probably not very important in natural selection and certainly it is not the only consideration. There is a danger in mutation. Most mutations do not help the species survive. A species and an environment exist in balance with each other. Populations simply adapt to their current surroundings and the changes to them; they do not necessarily become better in any absolute sense over time. A change in the environment may require a change in the species in order for it to survive. But if a mutation spreads too quickly across an entire species it may prove maladaptive to the species if the environment undergoes a further change. More diversity in mutations and hence change is probably better than speed in a mutation becoming widespread in a species.

Related to this question, an important principle of natural selection is that a trait that is successful at one time may be unsuccessful at another. This principle was demonstrated by the classic experiments by Drs C. Paquin and J. Adams.[37] They developed a yeast culture and maintained it for many generations. Every so often a mutation would appear that allowed its bearer to reproduce better than its contemporaries. These mutant strains would crowd out the formerly dominant strains. Samples of the most successful strains from the culture were taken at various times. In later competition experiments, each strain would out-compete the immediate previous dominant type in a culture. But interestingly, some earlier strains could out-compete strains that arose later in the experiment. Competitive ability of a strain was always better than its previous type. Yet competitiveness in a general sense was not increasing. The success of any organism depends on the traits of its contemporaries. There is likely no optimal design or strategy for most traits, only ones based on chance such as the competition and the environment.[38, 39, 40]

ARE FEWER BOY BABIES BORN DURING HARD TIMES?
(Asked by Jack O'Connor of Dublin, Ireland)

The sex ratio of boy and girl births is affected by severe stressors such as earthquakes, tsunamis, environmental toxin contamination, political and social upheavals, and even serious downturns in the economy. When society gets such a severe shock, the number of boy births falls and the number of girl births rises. A reduced number of boys are conceived and a higher number of boys die before birth. The theory behind this is that natural selection favours female births when times are hard, because on average females have a better chance of mating successfully than males. This principle seems to be followed in animals.

In 2003, evidence emerged that the same principle applies to humans. An article by Dr Ralph Catalano[41] presented an examination of birth records from Germany.[42] They reveal that in 1991, immediately following German reunification, the ratio of boys to girls in the former East Germany dropped to its lowest point since 1946, but then bounced back a year later. In the rest of Germany, where conditions were more stable, the boy/girl ratio was unaffected.

In 2006, Dr Catalano showed that the same odd occurrence happened in New York City after 9/11. The sex ratio of male births in New York City dropped in the period 1 January to 28 January 2002.[43] In addition, Dr Catalano has put forward the view that a male's life expectancy is affected by whether or not he was born in stressful times. Dr Catalano and colleague Dr T. Bruckner evaluated two rival theories accounting for this reduced male foetal morbidity. The first is the 'damaged cohort' theory. This theory holds that a mother's response to the shocks of stressful times can trigger 'stress reactivity' in the foetus and thereby shorten the life span of males in utero. The second is the 'culled cohort' theory. This theory holds that shocks of stressful times induce spontaneous abortions of frail male foetuses, but hardy male foetuses survive. Drs Catalano and Bruckner examined data from several northern European nations and conclude that there is more support for the 'culled cohort' theory.[44, 45, 46]

IS THERE A DIFFERENCE BETWEEN CRO-MAGNON MAN AND MODERN MAN?
(Asked by Nikki Jackson of Mosman, New South Wales)

Cro-Magnon humans are regarded as the earliest form of modern humans. We are their direct descendants. Cro-Magnons lived from about 40,000 to 10,000 years ago. If you saw a Cro-Magnon walking down the street, washed, dressed, shaved and with a haircut, they would be indistinguishable from modern humans. The only differences would be: 1) a slightly larger and more muscular body, and 2) a somewhat larger skull and brain. There is more variation among people you would see on the street now, so Mr or Ms Cro-Magnon would probably escape your attention.[47]

WILL ROBOTS EVER BECOME JUST LIKE HUMANS?
(Asked by Chuck Schroeter of Seattle, Washington, USA)

We are getting very close to building an almost human-like robot, however, there will never be a *completely* human-like robot since robots will never be able to biologically reproduce. Other than this, theoretically, robots could look, act, think and feel like humans in every way. The technical problems that need to be overcome in achieving this are being addressed one by one.

The development of humanoid robots today focuses on three major areas: 1) control of manipulators (what it can do); 2) biped locomotion (how it can walk); and 3) interaction with humans.

◆

The movements of robots have previously been awkward and non-human like. In 2006, the 'Flexible Spine Belly-Dancing Humanoid', a robot that can belly dance just as well as a human, was developed by Dr Jimmy Or.[48, 49, 50]

◆

Human-like 'walking in a straight line to making a turn has been achieved with the latest humanoid robot ASIMO' according to Drs M. Hirose and K. Ogawa.[51, 52]

◆

Dr A. Arita and colleagues[53] found that humanoid robots are now so well developed that they are regarded as human beings by 10-month-old infants in laboratory experiments.[54]

◆

The latest humanoid robot of Dr Hiroshi Ishiguro,[55] the Geminoid H1-1, is a near perfect replica of its creator — down to the frown in his face.[56]

◆

People find robots more likeable the more human they look, but only up to a point — once they become too lifelike they become frightening.[57]

◆

The human-sized, humanoid robot 'H7', invented by Dr K. Nishiwaki and colleagues,[58] is designed for autonomous walking, performing tasks and interacting with humans in an indoor environment.[59]

◆

French researchers led by Dr Alain Cardon[60] 'propose to develop a computable transposition of the consciousness concepts into artificial brains, able to express emotions and consciousness facts'.[61]

◆

Skin being developed at MIT will enable robotic hands to sense when something is slipping through their fingers — and to react to stop it.[62, 63]

HOW OLD IS MY BODY IF MY CELLS KEEP RENEWING THEMSELVES?
(Asked by Jo Hopkins of Belmont, California, USA)

About a century ago scientists first discovered that most of our brain cells formed during foetal development stay with us throughout life. But this discovery stimulated other scientists to study the age of cells throughout the human body. If we look at the adult human body from head to toe at age 40, the list goes something like this:

● Brain cells of the cerebral cortex (the grey matter) are with you from birth.
● Brain cells of the visual cortex (the array of cells in the front of the brain used for vision) are with you from birth.
● Brain cells of the cerebellum (the structures at the base of the brain) are slightly younger than you are.
● Intercostal muscle cells are about 15.1 years old.
● Gut lining cells are about 5 days old.
● Gut cells other than the lining are about 15.9 years old.
● Skin cells are about 14 days old.
● Red blood cells are about 120 days old.
● Bone cells are about 10 years old.
● We don't know precisely the average ages of eye-lens cells, heart cells, liver cells, pancreas cells, fat cells and bone marrow cells.[64]

HOW MANY CELLS ARE IN THE HUMAN BODY?
(Asked by Jeanne A. of Guadeloupe)

The answer depends on whether you mean types of cells or if you mean overall number of cells of any type. Most people are surprised to learn that it is not merely a question of counting cells or measuring the size of a person and making a calculation as to how many cells are in the human body. In fact, there are at least 210 different types of cells in the human body that qualify for individual names. Here are just a few of the categories of human cells: keratinising epithelial cells, cells of wet stratified barrier epithelia, epithelial cells specialised for exocrine secretion, cells specialised for secretion of hormones, sensory transducers, autonomic neurons, eye-lens cells, pigment cells, germ cells and so on.

The names of cell types continually change too. For example, modern immunology has shown that the old category of 'lymphocyte' includes more than 10 distinct cell types. Beyond this, a very large number of cells in the human body are bacterial cells. Do these cells count as being part of the human body or not?

The human body grows and then shrinks from conception through to late adulthood. So calculating the number of cells becomes more complicated depending on the age of the person. Of course, the size of the person is also an important consideration. All other things being equal, including age, a 1.83 m (6-foot) tall person would have more cells than a 1.52 m (5-foot) tall person. A 100 kg person would have more cells than a 60 kg person. With all of this complexity and much more not mentioned here, it is no wonder that 'how many cells are there in the human body?' is a very complicated question indeed. Dr Michael Onken[65] estimates the number of cells in the human body as being from 10 to 100 trillion. This is a huge range. Not exactly precise, but there it is.[66]

❖

If you calculated the DNA length for each person on Earth, it would stretch across the diameter of the solar system.

❖

From the moment of conception, you spend about half an hour as a single cell.

◆

Side by side, 2000 cells from the human body could cover about 2.5 cm^2.

HOW CAN HUMANS BE 'ENHANCED'?
(Asked by Rodney York of Halifax, Nova Scotia, Canada)

There are currently five general areas of experimental 'therapies' that hold the prospect for human 'enhancement', that is, 'improvement', as controversial as this might be.

1. **DNA** We may one day gain the power to insert new genes safely into various parts of the adult body and perhaps someday also into gametes and embryos. According to Moore's law, the performance of computing doubles by any measure you use every 18 months. This has been the case for the last 40 years resulting in computer power increasing more than 100 million times over that period. This is why your microwave oven today has more computer power than all the computer power existing in the world 50 years ago. DNA research developments occur very rapidly now. The rate of change is increasing dramatically too. It took 15 years to sequence the genome of HIV. The SARS genome was sequenced in 31 days.
2. **Drugs** We may one day gain the power to enhance physical performance using drugs such as safe steroids.
3. **Cognition** We may one day gain the power to alter safely cognition, including memory, mood, appetite, libido and attention, through psychoactive drugs. There are currently over 40 cognition-boosting drugs in development designed to improve wakefulness, memory, decision making, planning and other aspects of thinking. One of these drugs is modafinil. Modafinil is a 'wakefulness promoter' that was originally developed to help people suffering from narcolepsy and other sleep disorders. It was discovered that for those who have normal cognition, modafinil

improves not only planning but also decision making, verbal memory and visual memory. Dr Barbara Sakakian[67] claims that modafinil is the 'first true smart drug' to be developed. It will not be the last.[68]

4. **Implants** We may one day have the power to replace body parts with natural organs, mechanical organs, or tissues derived from stem cells, perhaps soon we will be able to rewire ourselves using computer chips implanted into the body and the brain.

5. **Life extension** We may one day have the power to prolong not just the average, but also the maximum human life expectancy. Dr Aubrey de Grey[69] predicts that in 20 to 30 years it will be possible to deliver radical increases in longevity largely by repairing cellular and molecular damage. Dr de Grey was quoted as saying, 'I think the first person to live to 1000 might be 60 already.'[70] No surprise, not all scientists agree.

◆

The most useful type of stem cell is the pluripotent stem cell. It can divide indefinitely and develop into a variety of cell types, including nerves, blood, bone marrow, muscle and internal organs.

◆

An infertile couple from Houston, Texas, turned to in vitro fertilisation. Two eggs were implanted in the woman's womb. Both of them split. She gave birth to two sets of identical twins at once.

27

Chapter 2
The Head

ISN'T THE HUMAN SKULL JUST ONE BONE?

(Asked by Jill Maynard of Greenacre, New South Wales)

It surprises many to learn that the human skull is not one solid bone as is usually thought. Nor does it consist of just two bones — the top of the head and everything underneath. The skull (cranium) is actually composed of 22 separate bones! There are 8 cranial bones around the brain and 14 facial and jawbones in the human skull. Just one of these bones moves — the jawbone (mandible).

In infants and very small children, the cranial bones are disconnected segments held together by connective tissue strips called sutures. At certain sites, these sutures are especially weak, creating the so-called 'soft spots' (fontanels) in an infant's head. The most prominent of these is a little further up from the infant's forehead. When growing is complete, the bones of the skull fuse together along the suture lines. These unions contain small amounts of fibrous connective tissue similar to those of the joints of arms and legs. Although the skull may structurally appear to be one piece when fully developed, it is still composed of separate bones.

Many fossil skeletal remains that anthropologists find often appear to have cracked or broken skulls. But these skulls are actually just missing some of their pieces. The softer connective tissue having decomposed, little support is left between the individual pieces in the skull. This causes them to fall out and perhaps get left behind over thousands and thousands of years. Has anyone seen my ...? Then again, let's not get a head of ourselves.[1]

◆

A human dies almost immediately once its head is severed. A cockroach will live 9 days without its head before it starves to death.

◆

Banging your head against a wall uses 628 kJ (150 calories) in 1 hour.

WHAT IS A NERVOUS FACIAL TIC?

A nervous tic is an involuntary, compulsive, sudden, rapid, recurrent, non-rhythmic, stereotyped motor movement or vocalisation. Usually the face and sometimes the shoulders are involved in a tic. Most people do not know that a tic can involve speech as well as body movement. When a tic produces movement it is a motor tic. When a tic produces vocalisation it is a phonic tic. Tics can be classified as simple or complex, and so are divided into four categories:

1. A **simple motor tic** is typically sudden, brief and meaningless. Significant eye blinking or shoulder shrugging are examples. But other simple motor tics can occur too. Among these are hand clapping, neck stretching, head, arm or leg jerks, mouth movements and other facial movements.
2. A **simple phonic tic** is typically any noise such as throat-clearing, coughing or sniffing.
3. A **complex motor tic** is typically longer and more purposeful in appearance. Pulling at clothes, touching things and people, and involuntary movements, such as copying the movements of someone else (*echopraxia*) are examples.
4. A **complex phonic tic** includes repeating words just spoken by someone (*echolalia*), repeating one's own previously spoken words (*palilalia*), repeating words after reading them (*lexilalia*), compulsive shouting (*klazomania*), and spontaneous uttering of socially objectionable words (*coprolalia*).

According to Dr A. Czaplinski and colleagues,[2] the most common psychiatric conditions associated with tics are personality disorders, obsessive compulsive disorder, self-destructive behaviour and attention-deficit hyperactivity disorder. All forms of tics may be exacerbated by anger or stress. However, they are usually much less frequent during sleep. Perhaps the most famous tic disorder is Tourette's syndrome.[3]

◆

We use over 70 muscles of our face and neck to say 1 word.

WHAT CAUSES THE EYES TO TWITCH?

(Asked by Karen Watterson of Apple Valley, California, USA)

Most people do not realise that the small twitches of the upper or lower eyelids are usually not tics. They are called fasciculations or twitches. A fasciculation is a small local contraction of muscles visible through the skin indicating a spontaneous discharge of a number of fibres innervated by a single motor nerve filament. Strictly speaking, a tic involves an entire muscle, so a fasciculation may not quite qualify. What's the real difference between a tic and a twitch? Only your neurologist knows for sure.

◆

Twitching can occur in any voluntary muscle group but is most common in the eyelids, arms, legs and feet. Even the tongue may be affected. Twitching may be occasional or continuous. Interestingly, any intentional movement of the muscle causes fasciculation to cease immediately. They come back as soon as the intentional movement ceases.

◆

The pupil of the eye expands as much as 45 per cent when a person looks at something pleasing.

◆

In survey after survey around the world, the most common response from women to: 'What feature of your face would you most want to change?' is 'My eyes, I'd like them bigger'.

WHY DO WE FIND SOME FACES ATTRACTIVE AND NOT OTHERS?
(Asked by Anka Brautigam of Utrecht, The Netherlands)

It is 'The Case of the Fertile Face'. Traditional evolutionary biological theory suggests that we are attracted to certain faces and not others because we are programmed to desire a healthy mate who will help us bear healthy offspring. We are therefore drawn to the face of someone we think will make a successful sex partner rather than someone we think who will not. However this theory has not always been supported in laboratory experiments. Gender (male or female) and sexual orientation (heterosexual or homosexual) seem to play a role. For example, a heterosexual female brain may show signs of 'attraction' during a laboratory experiment in which she sees an attractive woman. Yet she knows she will never mate with or ever want to mate with the woman. So why does 'attraction' register?

The 'fertile face attraction theory' has been challenged by Dr Alumit Ishai,[4] who claims that male or female, gay or straight, we all recognise an attractive 'salient sexually-relevant face, irrespective of their reproductive fitness'. We do not need to desire to mate with someone to find them attractive.[5] In other words, mating is something, but it may not be everything!

Over the years, researchers have speculated that we tend to be attracted to the faces of the opposite sex who remind us of our parents. However, others claim that there is just as much evidence supporting the opposite. Some researchers have even found that we tend to be attracted to those faces that remind us of ourselves. This was the case in a 1998 experiment, with a team of researchers led by Dr David Perrett.[6] In their experiment, the Perrett team morphed a digitised photo of the subject's own face into a face of the opposite sex. The subject then had to select from a series of photos which one he or she found most attractive. According to the Perrett team, subjects always preferred the morphed version of their own face — and they never recognised it as their own.[7, 8]

◆

An article in *Human Nature* (2002) concludes, among other things, that chickens prefer beautiful people compared to ugly people and that 'the animals showed preferences for faces consistent with human sexual preferences (obtained from university students).'

LOVE AT FIRST STARE!

It has been found that simply staring into each other's eyes has a tremendous impact on falling in love. Dr Arthur Aron[9] and colleagues draw this conclusion from their simple experiment. The Aron team put subjects, strangers of the opposite sex, together in pairs for 90 minutes and asked them to discuss intimate details about themselves. The researchers then had the subjects stare into each other's eyes for 4 minutes without talking. Many of the subjects felt a deep attraction for their partner after the experiment. A few even ended up getting married within a year![10, 11]

◆

The human male often 'loses his head' in the emotional and physical ecstasy of mating. By contrast, the male praying mantis often cannot mate while its head is attached to its body. The female may initiate mating by ripping the male's head off or she may end mating by doing the same. Occasionally, the male escapes with his head intact.[12, 13]

STALKING IS ADDICTIVE!

Drs J.R. Meloy and H. Fisher have found that what happens inside the brain when someone is stalking another person parallels to what happens in the brain when someone is suffering from an addiction.[14, 15]

HOW CAN AFRICAN WOMEN CARRY HEAVY BASKETS ON THEIR HEADS?

(Asked by Tanjina Ahmad of Lakemba, New South Wales)

It's not the basket that's heavy, but what's inside it! Seriously though, this ability was studied by an international team of researchers more than a decade ago. Their findings were first published in 1995 with subsequent studies appearing since. But it is not just African women who can accomplish this seemingly incredible feat. Sherpa women of the Himalayas do it too — as *any* woman (or man) can — once you learn how.

The researchers led by Dr N.C. Heglund[16] concluded that African women have learned the secret of transporting heavy loads on their heads with a minimum of effort or stress. It was found that some Kenyan women of various tribes can carry up to 20 per cent of their own body weight without making *any* extra effort! But techniques for doing so differ slightly among tribes. For example, Luo women balance loads on the top of their heads unaided, while Kikuyu women support the load on their head with a strap across the forehead. In either case, this is far greater than what European women can do.

It was also found that carrying a heavy basket on the head is much more efficient than carrying a heavy backpack on the shoulders. Why? The Heglund team claims that the secret lies in the women's pendulum-like motion in their posture and gait. A pendulum has its maximum energy potential at its highest point. A perfect pendulum transforms all its energy into body (kinetic) energy which reaches a maximum at its lowest point. The more a weight is kept at this highest point, the less additional energy is needed to keep it there. Essentially, when you carry a heavy weight on your head, the energy transfer is almost 100 per cent, but when the same heavy weight is carried on the shoulders in a backpack, most humans transfer only 65 per cent of the energy as they walk. The body's muscles are forced to make up the energy difference and that causes fatigue. So it's all just a matter of using your head.[17]

WHY IS IT CALLED A 'HEADACHE' IF THE BRAIN FEELS NO PAIN?
(Asked by Paulo Belluomini of Turin, Italy)

It is true that the brain feels no pain. This is why brain surgery can sometimes be performed while the patient is conscious. A headache is misnamed. Headache pain generally originates from the nerves, muscles and tissues *outside* the brain and skull. There are three likely sources of headache pain. The most common is the tension headache or nervous headache caused by contractions of the muscles of the scalp and back of the neck. Migraine headaches are produced from the dilation and constrictions of blood vessels in the scalp, temples and face. Sinus headaches are caused by congestion and inflammation of the nasal sinuses. Contrary to popular belief, the eyes and eyestrain seldom cause headaches, except with glaucoma. Although uncommon, headaches that are caused by a problem inside the brain are usually more serious. Of course, see your family doctor if you have any concerns about any headache.[18]

WHO GETS MORE HEADACHES: WOMEN OR MEN?
(Asked by Paulo Belluomini of Turin, Italy)

According to the American Council for Headache Education, almost 95 per cent of women and 90 per cent of men have at least 1 headache per year.[19]

HOW CAN HEARING THUNDER CAUSE A HEADACHE THAT LASTS FOR DAYS?
(Asked by Paulo Belluomini of Turin, Italy)

The 'Thunderclap Headache', according to the famous Mayo Clinic in Rochester, Minnesota, is an unusual type of headache which occurs when someone hears thunder. It is a severe headache that reaches its maximum intensity in less than a minute and can last for up to 10 days![20, 21]

WHAT IS A 'BRAIN FREEZE' AND WHY DOES IT HURT SO MUCH?

(Asked by Michael Friesen of Singapore)

The 'brain freeze' is another name for 'the ice cream headache'. The causes of this condition are eating ice cream or gulping a cold drink too quickly. Inhaling very cold air can cause a 'brain freeze' too. This type of headache is in the form of sharp, stabbing pain in the forehead, temples and around the eyes, and is intensely painful. The pain peaks at about 30 seconds or less after it begins, then it is almost always gone in less than 2 minutes. Cold material moving across your palate and the back of the throat is what brings on this headache.

One possible mechanism is that this temporarily alters blood flow in the brain, causing the brief headache. According to the Mayo Clinic, one of the few good things about an ice cream headache is that it is often gone in the time it would take you to say its medical name — 'headache attributed to ingestion or inhalation of a cold stimulus'. The Mayo Clinic adds that 'you may be more susceptible to these if you're prone to migraines'.[22, 23]

WHAT IS THE 'CHINESE RESTAURANT HEADACHE'?

This is the unfortunate name for one of the symptoms of the allergic reaction to monosodium glutamate (MSG). MSG can induce a headache in people who are allergic to it. The association with Chinese restaurants comes from the fact that MSG is often an ingredient in Chinese food. According to Johns Hopkins Medicine, such a headache typically begins within 30 minutes of consuming MSG and consists of dull and constant pain that may be at the front of both sides of the head. The headache typically goes away within 72 hours of eating a food containing MSG.[24, 25]

WHY DOES A HANGOVER OCCUR THE MORNING AFTER AND NOT THE NIGHT BEFORE?

(Asked by Anna Ro of Newtown, New South Wales)

A 'hangover' is technically called a 'delayed alcohol-induced headache' or DAI headache. According to the Mayo Clinic, symptoms of a DAI headache include 'a pulsating pain that is felt in front and on both sides of your head. It may worsen when you move around.' A DAI headache from excessive drinking is due to the following: alcohol contains the chemical ethanol; ethanol causes blood vessels to expand; such an expansion can give you a headache. Ethanol also disrupts the body's water balance and causes dehydration. Dehydration contributes to headaches. Congeners are flavouring ingredients frequently added to various alcoholic beverages, particularly the darker coloured ones. Congeners also can cause headaches. Congeners in high amounts can even be toxic.

When alcohol is metabolised by the body, the blood becomes more acidic than normal — this is called acidosis and can also contribute to a headache. Alcohol ingestion can alter the normal daily rhythm of various body functions. After the assault upon the body from too much alcohol occurs, it takes time for the body to respond to the stress and eventually get back to normal, so a hangover is not immediate. And just because you may have a headache that's DAI doesn't mean you should risk a DUI.[26, 27]

WHAT IS A POST-LUMBAR PUNCTURE HEADACHE?

(Asked by Liz Downes of Las Vegas, Nevada, USA)

There are all sorts of unusual forms of headaches that most people have never heard about. The post-lumbar puncture headache is one of them. This form of headache occurs after one has a spinal tap (lumbar puncture). Headaches of this kind are often accompanied by stiffness in the neck, ringing in the ears, light sensitivity, nausea and hearing impairment. The headache generally develops within a week of the spinal tap and typically resolves within a week. Such a headache can be debilitating. A post-lumbar headache may start when one stands up and stop when one lies down. There is some debate as to what percentage of patients who have a spinal tap also develop a post-lumbar puncture headache. An article by medical journalist Heidi Moore estimated that the figure was '[n]early a third or more'.[28] A review by Dr R. Gaiser[29] suggests that the figure is 'approximately 50%'.[30] According to a team of doctors from Johns Hopkins University led by Dr C.L. Wu,[31] males are less likely to suffer post-lumbar puncture headaches than females.

The lumbar region is the lower back. Anatomically, it is located at the back between the thoracic vertebrae and the sacrum. The lumbar region is sometimes referred to 'inaccurately' as the loins. (Lumbar is from the Latin *lumbus* and means 'loins'.)[32, 33]

WHAT IS TETANUS?
(Asked by Belinda Smith of Columbus, Ohio, USA)

Tetanus is an acute and often fatal infectious disease caused by the anaerobic, spore-forming bacillus *Clostridium tetani*. This bacterium is common and widespread throughout the world. It can be found in soil, dust, faeces and even on the surface of the human skin. The bacterium can only reproduce in the absence of oxygen. Thus, a deep wound becomes a good breeding ground for this bacterium. The bacteria usually enter the body through a contaminated puncture wound. Such wounds can be caused by metal nails, wood splinters and insect bites. But the bacteria can also gain entry into the body via the site of a burn, frostbite, surgery, cutaneous ulcer or injection. An injury that creates dead skin can lead to tetanus. A mother can get tetanus from the uterus after birth. A special kind of tetanus (*tetani* neonatorium) occurs in newborn babies from the site of the umbilical cord. The symptoms of tetanus can be severe. Once the bacteria get underneath the skin, they produce harmful toxins that attack the central nervous system. This results in uncontrollable muscle spasms and muscle rigidity throughout the body. The muscle spasms often look like unnatural stretches, hence the name tetanus. Tetanus comes from the Greek *tetanos* which means 'to stretch'. Tetanus strikes about 100 people in the US each year. Approximately 25 per cent of these people die of the disease.

WHY IS TETANUS CALLED 'LOCKJAW'?
(Asked by Belinda Smith of Columbus, Ohio, USA)

The uncontrollable muscle spasms and muscle rigidity that are the most characteristic symptoms of tetanus very frequently affect the face. The mouth does not open and the jaw seems as if it were made of stone, or 'locked shut', hence the rather apt description of 'lockjaw'.

CAN YOU GET TETANUS FROM A RUSTY NAIL?
(Asked by Belinda Smith of Columbus, Ohio, USA)

Whether it is rusty or not, any object that punctures or damages the skin can lead to tetanus. According to Dr Andrew Lloyd,[34] 'If you get a deep wound in a dirty environment, dirt contaminated with *Clostridium tetani* could enter the wound. The fact that the nail is rusty has no effect on whether or not tetanus develops.'[35]

◆

Physiological tetanus is a state of sustained muscular contraction without periods of relaxation caused by repetitive stimulation of the central nervous system at frequencies so high that individual muscle twitches are fused and cannot be distinguished from one another.

◆

Tetany is the 'hyperexcitability' of nerves and muscles. Tetany can be brought about by many things including a severe vitamin D deficiency.

◆

Historically, tetanus ravaged humans particularly soldiers during warfare. According to Dr J.D. Grabenstein[36] and colleagues, anti-tetanus vaccinations were introduced in 1933. During World War II there were only 12 reported tetanus cases among the 12 million US uniformed armed forces.[37, 38]

WHAT IS THE THYMUS?

(Asked by Ann Carter of Waukegan, Illinois, USA)

If the thymus was a B-grade movie, it could be *The Incredible Shrinking Organ*. The thymus is a lymphoid organ near the base of the neck atop the breastbone. It is a yellowish-grey or pinkish-grey blob of tissue that is, at its largest, about the size of a wristwatch. Like the brain, the thymus consists of two lobes.

Thymus is from the Greek *thymos* meaning 'a warty projection'. Some say the thymus resembles the leaves of a bunch of the minty herb thyme. At birth, the original weight of the thymus is about 18–19 g. As the baby grows into a child and the child into an adolescent, the thymus grows to about 113–114 g at puberty. But then something pretty amazing occurs. The thymus starts to shrink. The shrinkage is caused by the increased circulating level of sex hormones in the body. This is why when chemical or physical castration occurs to an individual, the thymus increases in size and activity. The thymus reduces in size until it is about 3.1 g at age 50. It is then smaller than it was at birth. Its function atrophies too.

Not long ago the thymus gland was considered to be a non-productive evolutionary appendage with no known function. But now we know differently. The role of the thymus in human health is great. From the late stages of gestation to the early stages of puberty, it is the chief defence mechanism against body infection. It is at this time that most of the T-cells (which stands for thymus cells) are formed and these will be carried by an individual for the rest of their lives. The thymus produces lymphocytes and supplies a hormonal stimulus to prod into action the spleen, the lymphatic system and other organs. If there's a possibility of infection, no matter how small, the lymphocytes produced by the thymus pour out the antibodies to attack and kill the invaders. The invaders may be bacteria, viruses, fungi or foreign tissue. There are millions of different antibodies, a special one for each disease, which come forth to fight off disease. At about age 50, the immune response slows, the thymus no longer produces lymphocytes and the defence mechanism begins to break down. This is attributed to the shrivelling up and almost complete disappearance of the thymus — the incredible shrinking organ.

WHAT IS THE THYROID?

(Asked by Ann Carter of Waukegan, Illinois, USA)

If the thyroid was a B-grade movie, it could be *The Vampire Bat Within*. The thyroid is a large endocrine gland. It produces hormones that regulate and affect growth and many other systems of the body. The thyroid looks like a small animal with wings resembling bat wings, or a large butterfly, or a little bow tie. The thyroid is located in the lower neck — below the voice box (larynx) and above the collarbones (clavicles). It is all around your Adam's apple. This is just about where a vampire would bite you or where a bow tie would be worn. Like the brain and the thymus, the thyroid consists of two lobes. The two lobes are connected by an isthmus, hence the similarity to a bow tie or a butterfly.

Thyroid is from the Greek *thyreos* meaning 'shield-shaped' or 'oblong door-like object'. Some have said that the thyroid resembles a shield used in ancient battles. The thyroid is a yellowish-white blob of tissue that varies in size. An overactive thyroid is called hyperthyroidism. An underactive thyroid is called hypothyroidism. Normally, you cannot feel or see the thyroid in your neck, as it weighs only about 10–20 g. But when it seriously enlarges and forms a goitre, you can easily feel and see it. Some human goitres are the size of a football. The human body needs iodine to keep the thyroid healthy. Twenty-five per cent of the body's iodide ions are located in the thyroid — as tiny as it is. The mental condition of cretinism is a deficiency of the thyroid.

DOES THE THYMUS OR THE THYROID PRODUCE THIAMINE?
(Asked by Ann Carter of Waukegan, Illinois, USA)

A frequently asked question is the relationship between the thymus, the thyroid and thiamine. The thymus is a lymphoid organ. The thyroid is an endocrine gland. Thiamine is another name for vitamin B1. The three sound as if they might have some relationship to each other but they don't — neither the thymus nor the thyroid produces thiamine.

◆

When thiamine was discovered by Umetaro Suzuki in 1910 it was called aberic acid. This is because the discovery took place while Suzuki was researching rice as a cure for the disease, Beriberi.

◆

Thiamine dissolves in water, but not in alcohol.

◆

Thyroid cancer has increased among those exposed to the Chernobyl nuclear power plant accident. This is according to a research team led by Dr A Prysyazhnyuk.[39] Their findings cover

360,000 residents, 50,000 evacuees and 60,000 workers who participated in the recovery. They write: 'In all groups under study a significant increase of thyroid cancer incidence rates has been registered. This increase appears to be associated, at least partly, with the fallout of radioiodine, and it was found not only in children, but also in adolescents and adults.'[40]

WHAT IS DANDRUFF?
(Asked by Amy Batisse of Channel 9's *Today*, Melbourne)

Everyone has a tiny bit of dandruff. Dandruff is merely skin that flakes free from the scalp. About 60 per cent of us have more or less normal amounts of it and about 40 per cent have either the occasional severe case of it or have chronic severe dandruff. The term dandruff is from two ancient Scottish words, *dandy* meaning 'dim' and *hurf* meaning 'scab'. Dandruff used to be called 'scurf' from the same *hurf* and was medically known as *Pityriasis capitis* which loosely translates as 'sorry condition of the head'.

A person has an average of 7.4 million skin cells per square centimetre of skin surface. Skin cells are born, they live and then they die. Every person must shed dead skin cells, including on the scalp. Hair holds back much of this normal shedding of dead skin cells — this is dandruff. Dry skin due to diabetes can also cause dandruff. Totally bald people do not have dandruff unless they suffer from some other medical condition such as seborrhoeic dermatitis, psoriasis or fungal infection.

Dandruff is also thought by some to be caused by an overgrowth of yeasts such as *Pityrosporum ovale* which live on normal skin. This overgrowth causes local irritation resulting in too many keratinocytes (hyper-proliferation of the cells) forming the outer layer of the skin. These form skin cell scales which accumulate and are eventually shed.

Anti-dandruff shampoos work in three ways:

- Ingredients such as coal tar are antikeratostatic, so they inhibit keratinocyte cell division and hence there are fewer such cells to shed.
- Detergents in the shampoo are keratolytic, so they break up accumulation of the skin cell scales.
- Finally, antifungal agents such as ketoconazole inhibit growth of the yeast itself. Other components such as selenium sulphide also inhibit yeast growth and therefore skin cell scales.

◆

Each hair strand is about 0.008 mm thick.

◆

Seborrhoeic dermatitis is not the same as dandruff. Seborrhoeic dermatitis is a form of inflammation of the skin (skin rash) that results from an overactivity of the sebaceous glands of the skin. According to Dr R.A. Schwartz and colleagues:[41] 'Seborrheic dermatitis affects the scalp, central face, and anterior chest. In adolescents and adults, it often presents as scalp scaling (dandruff).' They point out that stress can cause flare-ups and the skin scales are actually greasy and not dry.[42]

Chapter 3
The Eyes

HOW FAR CAN THE NAKED EYE SEE?
(Asked by Tommy Wooltorton of Bristol, UK)

The human eye can see an almost unlimited distance. Looking up into the sky on a very clear night, the Triangulum Galaxy can sometimes be seen. This is a distance of 3.14 million light years. The Andromeda Galaxy is also sometimes visible. This is a distance of 2.5 million light years. However, when we see these galaxies what we actually 'see' is something at a somewhat lesser distance. What we see is light from the faraway object rather than the object itself.

What happens is that light coming from the object strikes our optic nerve. The optic nerve signals the brain. The brain then interprets this message and forms an image. The brighter the object, the farther away it can be from us and still be seen. The star, Rigel, is many times brighter than another star, Alpha Centauri. Yet Alpha Centauri is many times closer. According to Dr Brent Archinal,[1] the human eye can see about 2500 stars in the clearest of night skies. But usually only about 1500 to 2000 stars can be seen due to weather, pollution and other obstructions, and fewer still in night skies of cities due to city lights. Five planets can be seen from Earth with the naked eye under the best of circumstances: Mercury, Venus, Mars, Jupiter and Saturn.[2]

CAN YOU REALLY SEE THE GREAT WALL OF CHINA FROM THE MOON?
(Asked by Tommy Wooltorton of Bristol, UK)

It is one of the greatest urban myths that astronauts can see the Great Wall of China from the surface of the Moon. They cannot. Astronaut

Michael Collins in his book *Liftoff* (1988) wrote that there is a false notion that the Great Wall of China is visible from the Moon. Collins orbited the Moon during the Apollo 11 mission in 1969. Apollo 11 astronaut Neil Armstrong has stated many times that the Great Wall is 'definitely not visible from the Moon'. Apollo 8 and 13 astronaut Jim Lovell made very careful observations and says that 'the claim is absurd'. Apollo 15 astronaut Jim Irwin has said that seeing the Great Wall from the Moon 'is out of the question'. Fellow astronaut William Pogue orbited the Earth in the Skylab Space Station during 1973 to 1974. The altitude was about 482.8 km above the Earth. Pogue wrote in his book, *How Do You Go to the Bathroom in Space?* (1991) that he *could* see the Great Wall of China from the space station — but he needed binoculars to do so. Astronaut Jay Apt orbited the Earth a total of 562 times on four Space Shuttle missions from 1991 to 1996. In 1996 he wrote that 'we look for the Great Wall of China. Although we can see things as small as airport runways, the Great Wall seems to be made largely of materials that have the same colour as the surrounding soil. Despite persistent stories that it can be seen from the Moon, the Great Wall is almost invisible from only 180 miles (290 km) up!'[3] The Great Wall of China is only about about 6 m in width. That is not a big target to see from the Moon — an average of 384,400 km away![4, 5, 6, 7]

WHY DO BABIES BLINK LESS OFTEN THAN ADULTS?
(Asked by Cade Stevens of Nashville, Tennessee, USA)

In normal circumstances, newborn babies blink at the rate of less than 2 times per minute. In childhood, the blink rate rises, and so by about age 14, the blink rate is about 10 times per minute. In adulthood, the blink rate remains at about 10 to 15 times per minute. This rate changes with attention, stress, excitement, eye irritation and amount of sleep. Generally, when one has had more sleep, one needs to blink less. There is widespread variation in blink rate among individuals.

There may be a genetic component in this, as well as a learned cultural component.

Blinking also may play an undetermined role in body language as it does in our non-human primate cousins. The main physiological purpose of blinking is to spread tears over the surface of the eyes. According to Dr Samuel Salamon,[8] it is puzzling as to why babies don't suffer from dry eyes due to their lack of blinking. It could be that since babies sleep so much compared with adults and thus spend so much time with their eyes shut, perhaps dry eyes are less of a problem for them. Dr Salamon also points out that a baby's eyes have smaller fissures compared with an adult's. That is, much less of the front of a baby's eye is exposed to the outside world — and to its dirt, dust and brightness — due to the shape of the infant's skull. As the baby's eyelid openings are smaller in relation to the eye compared with the eyelid openings of an adult, a baby's eyes may need less lubrication.[9]

◆

Babies do not manufacture tears during their first month of life.

◆

A dry cornea can affect the outcome of today's high-tech eye scans using tomography (OCT). According to Dr D.M. Stein and colleagues,[10] corneal dryness affects OCT scan quality and measured nerve fibre layer thickness after only a short exposure time. What's to be done? They recommend 'to instruct those who are scanned to blink frequently or to instil artificial tears.'[11] Your mother was right: Rest your eyes from time to time.

WHY ARE SO MANY PEOPLE TODAY SHORT-SIGHTED?
(Asked by Alan Harper of Oakland, California, USA)

The eye receives rays of light and bends them so that an image is resolved on a small point of the retina. But things can go wrong. If the rays focus in front of the retina, the person has myopia (or short-sightedness) and suffers blurred vision of distant objects. But if the rays focus at a point behind the retina, the person has hyperopia (or long-sightedness) and suffers blurred vision of nearby objects.

According to Dr Stephen Miller,[12] 'the shape of the eyeball and the focusing power of the lens and cornea help determine focus, but the angle at which light rays hit the eye plays a role. Light comes into the eye from all directions. Rays entering the eye at an angle from above or below would tend to focus somewhere in front of or behind the centre of vision. Those rays coming in essentially perpendicular to the eye, on the other hand, would tend to be focused more directly on the retina, providing a clearer image of what one is looking at.'[13]

Myopia occurs in at least 7 different forms and at varying degrees of severity; it can first develop in infancy, youth or adulthood. The prevalence of myopia varies from country to country. Depending on how it is defined, myopia rates are as high as 70 to 90 per cent in Asia, 30 to 40 per cent in Europe and the US, but only 10 to 20 per cent in Africa. A 2005 study by Dr N.S. Logan and colleagues[14] found that slightly more than 50 per cent of UK first-year university students are myopic.[15] School myopia appears during childhood school years. This form of myopia is attributed to use of the eyes for close work at school. As humans use their eyes more and more in close activities (reading, computers, video games, television and so on) in our modern world, it's not surprising that so much myopia occurs.[16]

DOES A CROSS-EYED PERSON SEE DIFFERENTLY FROM SOMEONE WHO ISN'T?

(Asked by Michel Durinx of Leiden, The Netherlands)

The medical term for 'cross-eye' is strabismus; another term for it is heterotropia. But it goes by other popular names too, including 'wandering eye', 'squint eye' and 'walleye'. 'Walleye' is often thought of as the opposite of 'cross-eye'. 'Cross-eye' is when the eye points inwards. 'Walleye' is when the eye points outwards.

Strabismus is a chronic abnormality of the eye from the normal visual axes (ocular deviation). The eyes don't point in the same direction. *Strabismos* is Greek for 'squinting' and the type of strabismus is referred to as a tropia. *Troupe* is Greek for 'turning', and the prefix indicates the direction of the ocular deviation. So when one eye is rotated around its visual axis with respect to the other eye it is 'cyclotropia'. *Cyclo* is also Greek for 'turning'. When the ocular deviation is such that one eye turns inwards it is 'esotropia' and when outwards it is 'exotropia'. When the ocular deviation is such that one eye turns upwards it is 'hypertropia' and when it turns downwards it is 'hypotropia'.

Eye muscles (extraocular muscles) bring the gaze of each eye to the same point in space. Strabismus occurs when there is a lack of coordination of these muscles. This results in poor binocular vision that affects depth perception. Short- or far-sightedness can result because the brain cannot fuse the two different visual images into one. In infancy, congenital strabismus can cause amblyopia. This is a condition in which the brain ignores input from the ocular-deviating eye, although it is capable of normal sight. Amblyopia is also called 'lazy eye'. Sometimes when the bridge of an infant's nose is wide and flat and when there are skin folds in the corner of the eyes, there is an appearance of strabismus. This is called 'false strabismus' or 'pseudostrabismus'. As the child grows, the skull develops, resulting in the bridge of the nose narrowing and the folds in the corner of the eyes going away. There is no problem with vision. The test for this involves the doctor shining a light into the child's eyes. The light's reflection off the pupil should be in the same spot of each eye if there is no strabismus.[17]

WHAT'S WORSE FOR YOUR EYES: READING IN BAD LIGHT OR DOING CLOSE ACTIVITIES?
(Asked by Leo Sanualio of Greenacre, New South Wales)

Prolonged near activity is more hazardous to one's vision than dim light. This is according to Dr Michael Lawless.[18] He adds: 'Near activities refer to those done closer than arm's length, such as reading, playing video games, doing homework or staring at a computer screen.' Dr Lawless claims that Taiwan research has demonstrated a link between extensive periods of time doing near activities and the development of myopia in children.[19] According to a study by Dr M.A. Bullimore and colleagues,[20] 1 of the 4 major risk factors for developing adult myopia is 'a high proportion of time spent performing near tasks'.[21]

❖

If your legs worked as hard as your eyes do each day, they would have to walk about 80 km.

❖

Some researchers believe that men risk eye damage by wearing their ties too tightly.

IS A COMPLETELY ARTIFICIAL EYE CLOSE TO REALITY?
(Asked by Leo Sanualio of Greenacre, New South Wales)

A California researcher predicts that an artificial eye is 'in sight' in the near future. Dr Armand Tanguay Jr[22] is currently building the world's first implantable camera for the blind. He also predicts that by 2014 we will see the introduction of a 1000-electrode implant that will allow the blind to recognise faces and read half-inch type.[23, 24]

CAN BLIND PEOPLE SEE COLOUR BY TOUCH?
(Asked by Leo Sanualio of Greenacre, New South Wales)

German scientists are baffled by a blind woman proving on TV that she can 'see' colours by touch — at least enough to tell one colour from another. Gabriele Simon, 48, from Wallenhorst, Germany, revealed her ability on Germany's most popular TV show, *Wetten Dass*. She uses her fingertips to recognise the different colours of various T-shirts and blouses while blindfolded. Ms Simon says: 'It took me 20 years to master this skill. It is a combination of pure learning and concentration. This ability really gives me more independence, as I don't need to ask my mother about what to wear anymore.'[25, 26]

WHAT IS 'INATTENTIONAL BLINDNESS'?

Do you ever wonder how a magician is able to fool you with a trick, as they say, 'right before your very eyes'? Concentrate as much as you want, you cannot see the sleight of hand. Most of us believe that when we are looking at something, especially when we are *really* concentrating, we see everything that is important to see. But this is not true, and we have many behavioural studies to prove this. As Dr Daniel J. Simons[27] writes: 'Although we intuitively believe that salient or distinctive objects will capture our attention, surprisingly often they do not. For example, drivers may fail to notice another car when trying to turn or a person may fail to see a friend in a cinema when looking for an empty seat, even if the friend is waving.'[28] This behavioural phenomenon is called 'inattentional blindness'. Some of the findings from experiments with inattentional blindness have been fascinating:

- When subjects are watching the movement of blue balls across the screen and trying to predict their direction, and then the balls suddenly change to green, 88 per cent of subjects fail to notice the colour change.
- When subjects are watching a video of two basketball teams passing the ball back and forth and then the uniforms of one of the teams suddenly change colour, 1 out of 4 subjects fails to notice the change.

- When subjects are watching a video of a basketball game and a woman carrying an umbrella suddenly appears on the court amid the players and the action and remains for as long as 4 seconds, again, 1 out of 4 subjects fails to notice her.
- When subjects are viewing 2 crosses on a screen while trying to judge which is longer over several trials, and on the fourth trial 1 of the crosses suddenly becomes a rectangle, yet again, 1 out of 4 subjects fails to notice the change.

Some theories that account for aspects of inattentional blindness include:

- Dr S.B. Most and colleagues[29] wrote in 2005 that 'the most influential factor' that affects how well one notices is 'a person's own attentional goals'. So, as the theory goes, when you are watching a football game and vitally interested in the game's outcome, you are less likely to notice the bikini-clad girl standing on the sidelines holding a big sign in the shape of a heart.[30]
- Based on their experiments in 2006, Drs Mika Koivisto and Antti Revonsuo[31] theorise that the closer the change is to what we are already seeing the *more* we will notice the change. It would seem that this is counter-intuitive and it should be just the opposite. But presto! There it is. That's what subjects reveal.[32] Unlike magicians, nothing is up their sleeve![33]

WHY DOES A CAMERA FLASH BLIND YOUR EYES TEMPORARILY?
(Asked by Jo Walker of Longueville, New South Wales)

The camera flash makes you go temporarily blind since such a bright light can temporarily 'blanch' the light-sensitive rods and cones in the retina. After vision returns, the flash can also affect the perception of colour for a time. The same phenomenon happens to a lesser extent when you come out of a dark movie theatre into

51

bright sunlight. It takes a few seconds or even a few minutes to adjust. The eye has a remarkable ability to adapt to varying degrees of light. According to Dr Kenton McWilliams,[34] it takes the average human eye about 25 minutes to completely adjust to major changes in light.[35]

WHY DO I SEE MORE CLEARLY UNDERWATER WHEN I WEAR GOGGLES?
(Asked by Jo Walker of Longueville, New South Wales)

Ever notice that a spoon immersed in a glass of water appears bent? The same effect happens when you try to see underwater. When light moves from one medium (air) to another medium (water) it changes speed. Light travels more slowly through water than it does through air. As a result, the beam of light is bent. This is called refraction. The amount of refraction depends on the ratio of the speed of light through each medium.

The human eye is delicately balanced to make sure that light coming in to it through the pupil is perfectly focused onto the retina at the back of the eye. It is optimised for light coming from the medium of air not water because humans live in air more than they do in water. However, when our eyes are underwater, the light is bent by a different amount, so the light is not correctly focused. When you wear goggles underwater, you restore the eye/air medium interface such that normal sight is resumed. This optic phenomenon of light bending when it goes through different media is used to our advantage in spectacles and contact lenses. The lenses bend the light to correct the imperfect vision. Of course, focusing is not entirely about light. The focusing ability of the eye is also affected by the ciliary muscles of the eye located around the eyelids and eyelashes.[36]

ARE THERE PEOPLE WHO CAN SEE BETTER UNDERWATER THAN THEY CAN ON LAND?

As with almost everything about being a human being, there is variation among peoples of the world in the ability to see underwater. There are people in the world whose eyes are definitely more conducive to functioning underwater. Small nomadic tribes called 'sea gypsies' have lived for centuries among the islands of Southeast Asia. The children of these tribes regularly dive deep into the sea to collect food from the ocean bottom. Clams, sea urchins and other food stuffs are important parts of the diet and are depended on for survival.

One such sea gypsy tribe is the Moken, who live in the archipelago along the west coasts of Burma and Thailand. Moken children are famous for their aquatic skills. Research has discovered that they also have extraordinary underwater vision — at least twice as good as that of Europeans. A team of researchers led by Dr Anna Gislen[37] studied Moken children and reported their results in 2004 and again in 2006.[38, 39] The Gislen team found that when Moken children are diving they routinely and involuntarily constrict their pupils and change their visual focus (accommodation) in ways to better find food. Although some portion of this adaptation is acquired through use, some suggest that over centuries of diving for food genetic variations have further improved the abnormally acute underwater vision of Moken children. Can European children somehow learn how to obtain Moken-like 'water eyes'? Dr Gislen reports that through practice European children can develop better underwater vision in a few months. However, they still will not equal Moken children in this ability no matter how long they try. Perhaps where it can be developed, such a skill must be developed early or not at all.[40]

WHAT IS RAPID EYE MOVEMENT SLEEP?

(Asked by Nicole Frieze of Leipzig, Germany)

Rapid eye movement (REM) sleep is a normal stage of sleep when, not surprisingly, rapid eye movements occurs. REM sleep was discovered by two pioneering sleep researchers Drs Eugene Aserinsky and Nathaniel Kleitman[41] who published an article describing their discovery in 1953.[42] REM sleep in adults occurs during about 20 to 25 per cent of sleep. It is common to wake up for a short time at the end of a REM period. Most of our clearest-recalled dreams occur during REM sleep. Newborns spend more than 80 per cent of sleep in REM.

There are several theories as to why humans experience REM sleep. First, it is believed that this facilitates dreaming. REM is one aspect of what is happening throughout the brain and the body during some phases of sleep to make dreaming happen. Second, it is believed that certain memories are consolidated during REM sleep. Without REM sleep, our memories would not function as well. Studies show that memories concerning procedures and space are reinforced during this sleep period. Third, during REM sleep the brain's monoamine receptors recover. Correct monoamine reception in the brain helps fight depression. Thus, without REM sleep we would be more prone to depression. Fourth, REM sleep helps the human brain to develop from newborn through to adolescent. After the brain is developed, REM sleep continues in order to make subtle repairs to the brain. Fifth, a theory in 1998 by Dr David Maurice[43] has little to do with the brain, but everything to do with the health of the eye. Dr Maurice suggests that REM sleep evolved to keep fresh oxygen washing over the eye. He wrote that the eyeball movements during this sleep period 'stir the aqueous humour behind the closed lids and so avoid the risk that its stagnation could cause corneal anoxia [oxygen starvation]'.[44] More recently, Dr Maurice's theory was given support by Drs F. Hoffman and G. Curio.[45] They write that REM sleep helps prevent corneal erosion. During the second half of a night's sleep, there is mechanical irritation to the eye. This helps explain why we often wake up with red eyes and often want to rub our eyes upon waking. Drs Hoffman and Curio add that REM sleep phases 'ameliorate the nocturnal nutritional deficiency of the healthy cornea by shaking the aqueous humour, which stagnates during sleep'.[46, 47]

WHAT DOES 'SCOTOMISATION' MEAN FROM *THE DA VINCI CODE*?
(Asked by Rita Hamblyn of New York, USA)

In the blockbuster film, *The Da Vinci Code* (2006), there is an important scene where Robert Langdon, Sophie Neveu and Sir Leigh Teabing discuss the possible hidden images in Leonardo da Vinci's masterpiece, 'The Last Supper'. Does the picture depict John the Apostle or Mary Magdalene? Are there signs that Mary and Jesus are married? Where in the picture is the Holy Grail? Is it there at all? At one point in this scene, Sir Leigh refers to scotomisation. He hardly explains its meaning. It is safe to say that scotomisation is not a word in the current vocabulary of most film goers. Yet understanding this word and the concept behind it is key to unlocking one of the mysteries within a mystery within a mystery: Why do people perceive differently what is in 'The Last Supper', the artistic and historical mysteries surrounding it, and the mystery of why there has been all the excitement about *The Da Vinci Code*.

Scotomisation is the psychological tendency in people to see what they want to see and not see what they don't want to see — in situations, in themselves, in anything, even in a painting — due to the psychological impact that seeing (or not seeing) would inflict. In this case, it is one of the most famous paintings of all time and an icon in the faith of millions of Christians. The emotional power of this is considerable. It is no wonder then that *The Da Vinci Code* has been so controversial throughout the world.

Perception involves seeing and processing information through the filter of our intellect and our emotions. That's why people often see the same thing differently. Scotomisation can be a false denial but also a false affirmation of our perceptions. The term used in behavioural science is borrowed from the science of optics and ophthalmology. Scotoma is from the Greek *skotos* which means 'to darken' and refers to a spot on the visual field in which vision is absent or deficient. The French psychiatrist René Laforgue is thought to be the first to have used the term in a psychiatric sense. In a letter to Sigmund Freud, dated 10 June 1925, Laforgue wrote that 'scotomisation corresponds to the wish that is infantile ... not to acknowledge the external world but to put the ego itself into its place ...'.[48] At the time, Laforgue was talking about denial and repression in schizophrenics, but the term can have a more general application.

Psychiatrist R.D. Laing describes scotomisation as a process of an individual psychologically denying the existence of anything that they see with their own eyes that they really don't want to see and hence don't want to believe. He wrote in *Interpersonal Perception* (1966) that scotomisation is 'our ability to develop selective blind spots regarding certain kinds of emotional or anxiety-producing events'.[49] So it may be a matter of faith with the evidence of *The Da Vinci Code*. The extent of the scotomisation factor in our everyday relationships, judgments, decisions and all of our behaviour is almost impossible to calculate and is thus unknown. While it is true to say 'Seeing is believing!', the questions become seeing whom and believing what?[50, 51]

◆

The average newborn's eyeball is about 18 mm in diameter, from front to back (axial length). It grows slightly to a length of approximately 19.5 mm as an infant. It continues to grow gradually to a length of about 24–25 mm in adulthood. A ping-pong ball is about 35–40 mm in diameter. This makes the average adult eyeball about two-thirds the size of a ping-pong ball.

◆

The eyeball is set in a protective cone-shaped cavity in the skull called the 'orbit' or 'socket'.[52, 53]

◆

Everyone is colour blind at birth. The late actor Paul Newman was colour blind throughout his life.

◆

On average, the human eye blinks 4,200,000 times a year. A blink lasts about 0.3 seconds.

◆

Our pupils dilate when we look at someone we love. They also dilate when we look at someone we hate.

Chapter 4
The Nose

ARE THERE PEOPLE WHO HAVE NO SENSE OF SMELL?
(Asked by Lucy Altmann of St Kilda, Victoria)

People who cannot smell suffer from some form of nasal dysfunction of which there are several types:

- Anosmia is the complete loss of the sense of smell. Congenital anosmia is rare. It can run in families. Traumatic anosmia can occur due to an injury. Viral anosmia can occur due to an infection.
- Hyposmia is the partial loss of the sense of smell.
- Parosmia is a distortion of the sense of smell. People with this dysfunction smell one smell and confuse it with another.
- Phantosmia is when smells are imagined.
- Presbyosmia is the decrease in the sense of smell due to ageing.

◆

According to the website of the University of California, the nose can be affected by stress. This often presents as sinus headaches and/or as a runny nose. The condition is called vasomotor rhinitis.[1, 2]

IS THERE ANY EVOLUTIONARY ADVANTAGE IN SNORING?
(Asked by John Edwards of Hitchin, UK)

The topic of snoring has been addressed many times and in many ways. Several points that could be made about snoring:

- The reason that we snore more in old age is that the throat muscles involved in preventing snoring become somewhat weaker and more flaccid with age.

- Fatter people tend to snore more because fat deposits accumulate in the tissues of the airways. This makes the tissues heavier and causes them to block more of the normal line of airflow.
- An estimated 45 per cent of people snore from time to time and 25 per cent are habitual snorers.
- You snore more when sleeping on your back because in the supine position gravity causes the tongue to fall backwards somewhat. This can narrow the airways and partially block airflow.

What science has not answered is the puzzling question of what could be the evolutionary advantage in making a sound while snoring? Making noises (often very loud ones) while sleeping would seem only to advertise the fact that one is sleeping — and thus being more vulnerable to harm from other humans or predators. Making a sound while snoring would seem to provide no advantage and at least one disadvantage for survival. So how has this noisy behaviour survived natural selection? The only possible answer is speculative. Early humans slept in groups. Predators often hunted at night. It would seem to possibly be an advantage to the group if one or more members remained awake for good portions of the night due to someone else snoring. They would be able to wake the rest of the group and warn of possible dangers. Far fetched? Anyone care to add to this?[3, 4]

DOES YOUR NOSE SMELL WHAT YOUR EYES SEE?
(Asked by Charli Tricase of Amherst, Massachusetts, USA)

Research suggests that the nose does indeed smell what the eyes see. The human sense of smell is unreliable acting by itself. It needs visual cues to be more accurate. The underlying brain network involved in this olfactory (smell) and visual (sight) process has been studied by two London researchers. They administered a number of tests to subjects where the researchers watched the brain activity of subjects on an MRI as they perceived smell and sight stimuli. The researchers manipulated the stimuli in various combinations and watched the brain changes in subjects as they did so. The researchers, Drs J.A. Gottfried and R.J. Dolan[5] found that visual cues impact on the olfactory cues as registered in the frontal part of the brain's hippocampus and in the rostromedial orbitofrontal cortex.[6]

DOES THE COLOUR OF SOMETHING AFFECT HOW YOU SMELL IT?
(Asked by Charli Tricase of Amherst, Massachusetts, USA)

Colour has a profound effect on the perception of smells. For example, strawberry-flavoured drinks are judged to smell more pleasant when they are coloured red than when coloured green. And descriptions of the 'nose' of a wine are dramatically influenced by its colour. This is according to Dr R.A. Osterbauer and colleagues[7] reporting on their experiments with MRI technology. They found that colour cues affect olfaction as seen in changes in brain activity observed in the brain's caudal regions of the orbitofrontal cortex. In their experiments, laboratory subjects were given various combinations of smells — lemon, strawberry, spearmint or caramel — and colours — yellow, red, turquoise or brown. When a colour and smell matched expectations, such as a yellow colour together with a lemon odour, there was more activity in brain regions that process olfactory information than when the smell was given alone. Mismatches of colour and smell produced less brain activity. So actions such as adding red colouring to white wine can alter how a person perceives the wine's odour — and we can watch the brain doing this.[8]

DOES YOUR MOOD AFFECT YOUR ABILITY TO SMELL?
(Asked by Charli Tricase of Amherst, Massachusetts, USA)

French scientists have found that your mood does indeed affect your ability to smell. Dr S. Lombion-Pouthier and colleagues[9] presented evidence to support this in 2006. In their study of depression-prone people, they discovered that depressives have deficient smell sensitivity, a poorer ability to detect smells and a greater tendency to over-estimate the pleasantness of smells. The French researchers also showed that alcoholic, drug-addicted and anorexic people have impairments in their sense of smell.[10]

DOES SMOKING AFFECT YOUR ABILITY TO SMELL?
(Asked by Charli Tricase of Amherst, Massachusetts, USA)

Japanese scientists have discovered the extent to which smoking affects one's ability to smell when they studied patients who had undergone surgery for sinus problems. The researchers led by Dr K. Sugiyama[11] found that smoking-induced hyposmia (loss of the ability to smell) was more pronounced in patients who had been smoking for many years. In general, the greater the number of years of smoking the greater is the loss of smell.[12, 13]

WHY DOESN'T THE SMELL OF MY BABY'S DIRTY NAPPY BOTHER ME?
(Asked by Charli Tricase of Amherst, Massachusetts, USA)

Studies show that mothers are consistently less disgusted by the smell of their own baby's dirty nappy compared to the smell of dirty nappies of other babies. In a study headed by Dr Betty Repacholi,[14] mothers were given pairs of unlabelled nappies to smell. Some were from their own baby, others were from the baby of someone else. The mothers consistently judged the smell of their own baby's faeces as 'less revolting'. Why does nature seem to build this 'less revolting' mechanism into mothers? Mothers may merely become used to the smell of their own child's faeces. Or they may be picking up some olfactory cue of relatedness that we still do not entirely understand. But something more profound may be occurring too.[15]

According to Dr V. Curtis and colleagues,[16] disgust is a powerful human emotion that has been little studied until recently. They argue that 'the human disgust emotion may be an evolved response to objects in the environment that represent threats of infectious disease'.[17] So if disgust emerged as an emotion to keep us from getting close to unhygienic, rotten or potentially dangerous substances, being disgusted by the body products of our own babies would probably not help our species' survival at all.

WHY DOES THE SMELL OF A BREAST-FED BABY'S DIRTY NAPPY SMELL BETTER THAN THAT OF A BOTTLE-FED BABY?
(Asked by Charli Tricase of Amherst, Massachusetts, USA)

While cow's milk is perfect for a baby cow, human milk is perfect for the human baby. The mostly lactalbumin protein and fat in human breast milk are more easily digested by human babies than the mostly casein protein and fat in cow's milk. Cow's milk also contains more aluminium, cadmium, iron and manganese. Breast-fed babies have more watery movements than bottle-fed babies. Extra iron can also be somewhat constipating for the infant. This extra time in the digestive tract can increase smelliness. So the end product of this affects the smell, as more cow's milk is not digested compared to human milk.[18, 19]

WHY DO BABIES ALWAYS SEEM TO HAVE A RUNNY NOSE?
(Asked by Michael Woodhams of Palmerston North, New Zealand)

There are at least three reasons why infants and young children always seem to have a runny nose. First, they have more colds. Infants and young children have many upper respiratory tract infections due to a less developed immune system. This is why children often get so many colds when they start school. Close contact for the first time with other humans (and often infectious children at that!) exposes them to many viruses they have never encountered before. In the obverse, this is also why the elderly rarely have colds. Exposure to more viruses builds up immunity. Old people have had the exposure, young people have not. Second, infants and young children may not always keep as warm or as cool as they should. Thus, they may be more susceptible to vasometer rhinitis. This occurs when there is a change in temperature that causes swelling in the tiny blood vessels in the mucous membrane linings of the nose and produces a runny nose. Third, an infant or young child may not have fully developed sinuses, which could cause the nose to run more often.

According to Dr Vincent Iannelli, author of *The Everything Father's First Year Book* (2005), a baby's sinuses are not well developed. However, it is a myth that a baby has no sinuses at all. In fact, newborns have very small maxillary and ethmoid sinuses. The maxillaries are under the cheeks while the ethmoids are higher up in the nasal cavity. They are so small that they cannot be seen in a normal x-ray until the child is 1 to 2 years old. Dr Iannelli points out that 'the frontal sinuses and the sphenoid sinuses don't begin to develop until a child's second year and can't be seen on an x-ray until the child is 5 to 6 years old. The sinuses continue to grow until your child is a teenager.'[20, 21, 22]

CAN YOU STRIKE A PERSON ON THE NOSE AND KILL THEM BY DRIVING CARTILAGE AND BONE INTO THEIR BRAIN?
(Asked by Kevin Reinelt of Vancouver, British Columbia, Canada)

Anatomically, the nose is the anterior extremity of the head containing the nostrils, olfactory cavities and the olfactory organ. The hard part of the nose consists of cartilage which is connective tissue containing collagen, proteoglycan and chondroitin sulphate. Cartilage is softer and more flexible than bone. But it is still hard enough that if a blow to the nose was struck with enough force and at the right angle to both shatter the cartilage and propel it up into the brain, it could do enough damage to kill you.

CAN YOU DIE FROM PICKING YOUR NOSE?

Physicians who specialise in the nose are called rhinologists. They have for many years claimed that there is little danger from nose-picking other than perhaps breaking the skin and therefore risking infection. Of course, if seen doing it too much, you may lose a lot of friends — or perhaps die of embarrassment.

◆

Did we need to have a study to prove this? An article in the *Journal of Clinical Psychology* (June 2001) draws the earth-shattering conclusion that, among other things, nose-picking is a common adolescent activity.[23, 24]

WHAT IS 'THE DANGER TRIANGLE OF THE FACE'?

It is possible for infections from the nasal area to fairly easily spread to the brain due to the nature of the blood circulation system around the human nose. The skull is very thin in this part of the face making an infection there potentially more likely to involve the brain. For these reasons, rhinologists have designated the area from the corners of the mouth to the bridge of the nose, including the nose and the upper jaw bone (maxilla) as 'the danger triangle of the face'.

DOES A LITTLE AREA OF THE NOSE CAUSE NOSE BLEEDS?

Within the nose, often the cause of a nose bleed is an inflammation of a little area near the nasal septum that has many small capillaries. The medical term for a nose bleed is epistaxis. That little area of the nose is called, wait for it, Little's Area.

WHY DOES PEPPER MAKE YOU SNEEZE?
(Asked by Jang Dae Kim of Seoul, South Korea)

The essential oils found in pepper cause you to sneeze. These oils are extracted from the berries of the pepper plant and are used to flavour a variety of foods. The chief chemical in the pepper plant that causes sneezing is piperine. Piperine is found in both black and white pepper. When it is smelled and tasted, piperine gives you that pleasant pepper smell aroma to the nose and also a slightly biting sensation to the mouth. But since piperine is powerful enough to affect the tongue, it is also powerful enough to affect the delicate membranes of the nose. That calls out the body's sneeze response. The body's sneeze response is programmed to protect the body's internal airways from outside irritants and contaminants. Contact with piperine triggers the response. Table pepper is ground finely. So the fact that piperine is perceived by the body as a 'dust' as well as a chemical irritant triggers the sneeze response when pepper is inhaled. The sensors involved in the sneeze response go into action and attempt to expel the irritant — out the same way it came in. The result is a sneeze.[25]

◆

The longest known sneezing spell lasted 978 days.

◆

If you sneeze too hard you can fracture a rib. If you try to suppress a sneeze you can rupture a blood vessel.

DO OUR TWO NOSTRILS SMELL IN THE SAME WAY?

It is not widely known, but our two nostrils do not smell in the same way. Although the difference isn't that great, it helps our brain better perceive the rich world of fragrances. At any one time, air flows through one nostril more quickly than it does through the other. This is due to a slight swelling in the mucous membranes in one side of the nose. Every few hours, the swelling moves from one nostril to the other. The 'fast airflow nostril' becomes the 'slow airflow nostril' and vice versa. The swelling of the mucous membranes affects how quickly a scent is drawn into that nostril and how it is perceived by the brain.

Another fact about nostril differences is that various types of scents are absorbed at different rates through the mucous membranes, so a different 'smell profile' for each nostril is created. A fragrance that is absorbed quickly in the fast-airflow nostril is easier to detect. A fragrance that is absorbed less quickly in the slow-airflow nostril is stronger after it is detected due to the added time devoted to better fragrance absorption. The existence of different 'smell profiles' for each nostril was confirmed some years ago by researchers headed by Dr Noam Sobel.[26] In one experiment, 20 subjects were asked to sniff a substance made up of equal parts of octane and L-carnove. Octane is absorbed slowly while L-carvone is absorbed quickly. The subjects rated from 1 to 100 per cent the proportions of each they detected through each nostril. The Sobel team also measured the rate of airflow through each nostril. As predicted, it was found that 17 of the 20 subjects, sniffing through the fast-airflow nostril, gave the impression of more L-carvone, while the slow-airflow nostril signalled more octane. They concluded that 'each nostril perceives the world differently. The brain is receiving two disparate images of the olfactory world.'[27] So nostrils work a little differently, but this little difference helps us smell a lot better.

DO PHEROMONES WORK IN HUMAN SEXUAL ATTRACTION?
(Asked by Lisa McMillan of East London, UK)

Many people have learned that insects and other animals are driven by pheromones, and that perfumes and colognes are supposed to help in courtship, so they want a shortcut for successful romance. However, most scientists would say that there is little evidence that humans rely very much on pheromones as a sexual attractant. Pheromones are special chemicals produced by animals that serve to direct behaviour, including sexual behaviour. In mating, other animals rely on their sense of smell much more than humans do. It is argued that humans have virtually lost the ability to be attracted by pheromones and hence pheromones are only minimally important in human sexuality, if at all.

Nevertheless, some scientists contend that a tiny sense organ in our nasal cavity, the Vomeronasal Organ (VNO) which is sometimes called Jacobson's Organ, is capable of detecting chemical sexual attractants passed unconsciously between people. The VNO is located in the vomer bone between the nose and the mouth. How it functions in humans is disputed. In animals, it is much clearer:

- Mice use the VNO to detect pheromones — vital in mouse mating.
- Cats use the VNO to detect nepetalactone. This is what gives them the high from catnip.
- Snakes use the VNO to sense prey by sticking out their forked-tongue and withdrawing it — touching the VNO in the process.
- Elephants stimulate themselves by transferring sensory-stimulating chemicals to their VNO via the tip of their trunks.

In humans, the VNO first appears during foetal development. Strangely, it then shrinks to almost nothing by the time of birth — we don't know why this happens. In adults, a small pit can be found in the nasal septum of some people, but not in all. Again, we don't know why. Some scientists think that this tiny remnant means that the VNO still can work — at least in some humans. But what the VNO can do is anyone's guess.[28, 29, 30]

WHY DOES GARLIC MAKE YOUR BREATH SMELL BADLY?
(Asked by Charli Tricase of Amherst, Massachusetts, USA)

Unlike other vegetables, garlic has a powerful antibiotic and antifungal compound call allicin. Allicin has some medicinal purposes and can even be used in treatments for such conditions as hardening of the arteries (arteriosclerosis) and obesity. Allicin dissolves fats and is an antioxidant. Garlic has been used as a folk herbal remedy for centuries. According to Dr D.M. Bautista and colleagues,[31] 'the molecular mechanisms underlying these effects [of garlic] remain unknown'.[32] Interestingly, allicin is not in garlic in its natural state. Allicin is produced only when the garlic clove is damaged such as through chopping or chewing. When damage occurs to the clove, alliinase and alliin, two chemicals existing in the undamaged garlic, act upon each other to create allicin. Garlic belongs to the Allium family of plants that produce organosulphur compounds. Allicin is one of these. It is very pungent, hence creating 'garlic breath' when eaten.

WHY CAN'T YOU SMELL SOMEONE ELSE'S GARLIC BREATH IF YOU'VE BEEN EATING GARLIC TOO?
(Asked by Charli Tricase of Amherst, Massachusetts, USA)

The powerful, pungent, organosulphorous allicin compound in garlic affects a person's sense of smell when it is eaten. Your sense of smell and sense of taste are somewhat overpowered. It is not just the smell of garlic on someone's breath that you can't smell very well, it is actually everything you try to smell and taste. Some people remark that their own body odour is also affected.

CAN EATING PARSLEY CURE GARLIC BREATH?
(Asked by Charli Tricase of Amherst, Massachusetts, USA)

Parsley (*Petroselinum crispum*) has long been valued as a folk medicine breath freshener due to its high concentration of chlorophyll. But eating parsley to cure garlic breath could only give temporary relief at best. The allicin in garlic is just too powerful. Green cardamom (*Amomum subulatum*) in the ginger family has also been suggested as a possible garlic-breath antidote. In the folk medicine tradition of India, cardamom is used to treat mouth, gum and teeth problems. Cardamom has a strong and aromatic fragrance, and it can remain in the mouth for a considerable period of time. But again, the allicin in garlic is just too powerful. According to Trevor Mendham, 'Unfortunately, there's no way of getting rid of the symptoms of garlic eating totally.'[33, 34]

DOES GARLIC CURE A COLD?

(Asked by Charli Tricase of Amherst, Massachusetts, USA)

Garlic (*Allium sativum*) has been used as a folk medicine for at least 4000 years. Cloves of garlic were found by archaeologists in the tombs of ancient Egyptian pharaohs. But we don't know what the garlic was used for, maybe to cure a cold. According to Dr Hans Wohlmuth,[35] garlic has antibacterial properties that are well documented. These properties are primarily ascribed to the compound allicin. Allicin is the strongly smelling organosulphur-containing compound that is released when fresh garlic is chopped or chewed. But the cold is caused by a virus not by bacteria. So the effectiveness of garlic in treating the common cold is perhaps best left to physicians rather than to an anthropologist.[36] Yet a study concluded that 'an allicin-containing supplement can prevent attack by the common cold virus'. The study was conducted by Dr P. Josling.[37] There were 146 subjects in the study who took a garlic supplement or a placebo over a 12-week period. It was found that the garlic treatment provided effective protection against the common cold for many if not most of the subjects in the study. In addition, those who took garlic and still caught a cold usually had a significantly shorter duration of symptoms than those who did not take garlic.[38] Although published in 2001, the study has not been replicated. Even if the results are seen again, perhaps in a larger study and over a longer period of time, experts still have to determine the optimum dosage of garlic as a treatment for the common cold.[39]

DOES GARLIC CURE DIABETES?
(Asked by Charli Tricase of Amherst, Massachusetts, USA)

Some doctors think garlic as a treatment for diabetes should be researched further. In one study, garlic has been shown to reduce diabetes symptoms when diabetes is artificially induced in laboratory mice. Based on their rodent results, Dr A. Eidi and colleagues[40] write that 'it is concluded that the plant [garlic] must be considered as [an] excellent candidate for future studies on diabetes mellitus'.[41] Let's hope that what comes from this study is not Mickey Mouse.

CAN GARLIC BURN YOUR SKIN?
(Asked by Charli Tricase of Amherst, Massachusetts, USA)

One of the little known effects of garlic is that it can cause contact dermatitis when applied to the skin. In fact, garlic 'can cause a severe dermal reaction and a deep chemical burn', according to Dr T. Friedman and colleagues.[42] This was reported in their study of patients at their hospital who were treated for 'self-inflicted lower extremity burns' by applying garlic to their bodies.[43] Why anyone would want to do this to themselves is quite another matter.

DO HUMANS HAVE A PIECE OF IRON IN THEIR NOSE THAT HELPS WITH A SENSE OF DIRECTION?
(Asked by Lee Staniforth of Manchester, UK)

Some years ago scientists at Caltech in Pasadena discovered that humans possess a tiny, shiny crystal of magnetite in the ethmoid bone, located between the eyes and just behind the nose. Magnetite is a magnetic mineral also possessed by homing pigeons, migratory salmon, dolphins, honeybees and bats. Indeed, some bacteria even contain strands of magnetite that function, according to Dr Charles

Walcott,[44] 'as tiny compass needles, allowing them [the bacteria] to orient themselves in the earth's magnetic field and swim down to their happy home in the mud'.[45] It seems that magnetite helps direction finding in animals and helps migratory species migrate successfully by allowing them to draw upon the Earth's magnetic field. But scientists are not sure how they do this. In any case, when it comes to humans, according to some experts, magnetite makes the ethmoid bone sensitive to the Earth's magnetic field and helps your sense of direction. Some, such as Dr Dennis J. Walmsley and W. Epps[46] as far back as 1987, have even suggested that this 'compass' was helpful in human evolution because it made migration and hunting easier.[47] Following this fascinating factoid, science journalist Marc McCutcheon entitled a book *The Compass in Your Nose and Other Astonishing Facts About Humans* (1989).[48, 49]

◆

An article in *Parisitology Today* (April 1996) concludes, among other things, that a particular species of female mosquito that can carry malaria is attracted equally to the smell of limburger cheese as she is to the smell of human feet.

◆

At best, a human being can smell only about 1/20 as well as a dog.

Chapter 5
The Ears

DO OUR EARS GROW LONGER AS WE AGE AND IF SO, WHY?
(Asked by Judith Berry of Staffordshire, UK)

As we see others age or as we see ourselves age we often notice that ears appear to get longer. But do they really? The scientific validity of this common observation has been challenged from time to time by those who maintain that ears do not really grow longer (or larger) with age — they only *look* as if they do and it is all just an illusion. It has been pointed out that as the body shrinks somewhat with age, the ears may appear to have grown longer (and larger), while actually staying the same size. So what does science say? In fact, our ears *do* grow longer with age. Indeed, they grow throughout our lives.

Drs L. Pelz and B. Stein[1] measured the ears of 1271 children and adolescents. They reported in *Padiatrie und Grenzgebiete* that ear length increases 'steadily and annually', but ear width remains the same.[2] Dr James Heathcote[3] studied 206 patients with the mean age of 53. He concludes that 'as we get older our ears get bigger (on average by 0.22 mm a year)'.[4] Dr Yashhiro Asai,[5] along with his colleagues, agreed with Dr Heathcote. Their study of 400 consecutive patients aged 20 and older concludes 'that ear length correlates significantly with age in Japanese people ...'.[6] Finally, Dr V.F. Ferrario and colleagues[7] present evidence that not only do ears get longer with age, but it happens to both women and men. Men's ears start out longer than women's and they stay that way. Why do ears grow longer with age? Gravity over time forces all body appendages to sag. The bane of human ageing: If it can sag, it will sag! Ears included.[8, 9, 10, 11]

◆

It takes only about 4 kg of pressure to rip off your ears.

WHICH CAME FIRST: HUMAN MUSIC OR HUMAN LANGUAGE?
(Asked by Joel Jamal of Liverpool, UK)

When you first think about it, most of us would probably say that human language must have evolved before human music because we do the first much more than we do the second. But research suggests otherwise. A widely held scientific view today is that music itself is a protolanguage — a transition form of communication somewhere between silence, the calls of other animals and true human verbal language with words. In fact, it was Charles Darwin who put forward this view. In the *Descent of Man* (1871), Darwin argued that music was an early form of communication that is obviously common to birds and other non-human animals. He further reasoned that music and song probably preceded the spoken word. Darwin considered language so complex that it must have evolved through one or more of our immediate human ancestors who would have therefore been a language 'transition species'. Such early human ancestors would have 'conversed' not in words per se but rather in the pitch, rhythm and melody of vocal sounds. One of the best candidates for this was the then newly discovered Neanderthals. The Neanderthal became extinct between 230,000 and 300,000 years ago. They were discovered in the 1850s, but it was speculation as Darwin had little solid evidence to back his view. Many anthropologists today believe that Neanderthals did in fact communicate with musical sounds.

According to Dr Steven Mithen,[12] 'Neanderthalese' was basically musical! Dr Mithen is one of the leading advocates of the 'Singing Neanderthals' view.[13] He believes that the vocal capabilities of

Neanderthals some 1.8 million years ago were little different from modern humans. But they show no evidence of having symbolic thought that is a requirement for spoken language. Therefore, Dr Mithen asks, 'If not for language, why else would their vocal tract have evolved?' His answer: singing! Maybe Neanderthal singing was without words initially, but Mithen adds that it was 'with all the adaptive functions that music still serves (and does so more effectively than words), building social bonds, improving mental and physical health, advertising one's self to the opposite sex'. It is a fascinating possibility, and indeed something to sing about![14]

HOW DID BEETHOVEN COMPOSE MUSIC IF HE WAS DEAF?
(Asked by Janet Noonan of Vancouver, British Columbia, Canada)

The great music genius Ludwig von Beethoven (1770–1827) was not born deaf. However, by 1801 he showed definite signs of deafness. Although he would retreat more and more into seclusion, for the next 13 years he continued performing in public concerts. Even as late as in 1822, he courageously attempted to conduct his opera 'Fidelio' while hearing nothing. This proved to be disastrous as the conductor and orchestra eventually fell out of synch.

Despite progressive hearing loss, Beethoven was extraordinarily creative. He was able to compose while deaf because he could still remember what each note sounded like and could imagine what his compositions would sound like to others when played. In 1817, Beethoven asked the famous Streicher piano maker company to adjust a piano to make it as loud as possible. He asked another piano maker, Graf, to make a sounding board which, when placed on the piano, helped to better conduct the sound. Beethoven also touched the strings of the piano with a wooden stick clenched between his teeth so that he could feel the vibrations in the absence of being able to hear them. Nothing, not even deafness, was going to stop Beethoven from composing![15]

WHAT CAUSED BEETHOVEN'S DEAFNESS?

(Asked by Janet Noonan of Vancouver, British Columbia, Canada)

In 1990, Dr P.D. Shearer[16] claimed that Beethoven suffered from Paget's disease. This conclusion was based on the fact that Beethoven's large head and broad brow are very characteristic of this disease. Abnormal bone growth was believed to have crushed his auditory nerves causing his deafness.[17] But now there is more to the answer. In 2005, Drs C.S. Karmody and E.S. Bachor[18] argued that Beethoven's hearing loss was the result of, wait for it, 'immunopathy' from inflammatory bowel disease. In this surprising conclusion, they contend that hearing loss can sometimes result from such an intestinal problem.[19]

❖

The US National Institutes of Health estimates that 28 million people in the US suffer from some level of hearing loss.

❖

The crying of a baby can damage your hearing. A baby can cry at levels up to 90 dB. Permanent damage can occur at 85 dB.

❖

Wearing headphones for only 1 hour will multiply the number of bacteria in your ears by 700 times.

HOW DOES SCIENCE EXPLAIN JOAN OF ARC'S 'VOICES'?

Joan of Arc (1412–1431) is the most famous female military hero in Europe, a national heroine in France and a Catholic saint. During the horrific Hundred Years War (1337–1453), Joan of Arc's legendary military feats led to the liberation of France from England. What made her accomplishments even more extraordinary was that, apart from being a woman, Joan of Arc was young, illiterate and a peasant. Any one of these factors would have normally ruled her out of any military leadership role. Shakespeare, Voltaire, Schiller, Verdi, Tchaikovsky, Twain, Shaw and Brecht all created works about her, and contemporary books, films and plays continue to portray her. Today, the fascination with Joan of Arc seems to be as strong as ever — more than 550 years after she was burned at the stake for heresy. Joan of Arc attributed her success to the hearing of 'voices', which led her to do the things she did and in the way that she did them. Some believe these 'voices' to be divinely inspired visions. But what are the scientific explanations for the 'voices' of Joan of Arc? There are at least 11 theories:

1. She suffered from Menière's disease. This is an ear problem that can cause dizziness and auditory confusion sometimes taking the form of sounds similar to 'voices'.

2. She had tinnitus, a very common inner ear problem resulting in hissing or buzzing sounds in the ear that can occasionally resemble 'voices'.

3. She was schizophrenic; auditory hallucinations are common in sufferers of schizophrenia.

4. She was a psychotic. Delusions and hallucinations are hallmarks of someone suffering from the severe mental disorder of psychosis.

5. She was a psychopath. If so, she was a pretty clever one at that. At her trial, she maintained her wits and poise through torturous days of hostile cross-examinations by skilled interrogators of the Inquisition. They constantly tried to trap her into an admission of guilt. Amazingly she held her own — not bad for an illiterate peasant girl.

6. She was hallucinating from accidentally eating contaminated grain. The fungus ergot grows on grains such as rye (*Secale*), barley (*Hordeum*) and wheat (*Triticium*). Poor storage of grain, a common occurrence in the Middle Ages, allowed ergot to thrive. Ergot was known as 'mad grain' and 'drunken rye' because of the hallucinations it caused. The psychoactive ingredient in ergot is a form of LSD (lysergic diethylamide), and LSD would easily account for the 'voices'.

7. She pretended to hear 'voices'. It is said that she faked mental illness in order to gain favour from a sympathetic King Charles VII of France. The king's own father suffered from mental illness and was aptly known as 'Charles the Mad'.

8. In 1986, Dr R.H. Ratnasuriya[20] argued that Joan of Arc suffered from a form of tuberculosis that produced a tumour in the temporal region of her brain. Such a tumour can produce auditory hallucinations.[21]

9. In response to this theory, later that year Dr D.A. Moore[22] dismissed the tuberculoma theory and instead offered one of his own. His view was that the 'voices' were only one of several symptoms Joan of Arc exhibited. She also suffered profound ill health from malnutrition (cachexia) and the absence of menstrual periods (amenorrhoea) also due to malnutrition. So Dr Moore concluded that Joan of Arc more than likely suffered from anorexia nervosa. If so, then 15th-century Joan of Arc seems closely linked to 21st-century women.[23] Most people today think that anorexia nervosa is a modern malady, but the disease was also at epidemic levels in Europe during the Middle Ages. As Dr Rudolph Bell[24] argues: self-starvation was one of the few ways that subjugated women could attempt to take back some of the control of their lives through controlling at least some part of themselves — their bodies.[25]

10. In 1991, Drs E. Foote-Smith and L. Bayne[26] presented the theory that Joan of Arc suffered from epilepsy. They wrote: 'We suggest, based on her own words and the contemporary descriptions of observers, that the source of her visions and convictions was in part ecstatic epileptic auras and that she joins the host of creative religious thinkers suspected or known to have [had] epilepsy, from St Paul and Mohammed to Dostoevsky, [and] who have changed western civilization.'[27]

11. A theory that is probably not the last was put forward in 2004. Dr Maggie Phillips[28] — author of *Healing the Divided Self* (1995) and *Finding the Energy to Heal* (2000) — suggests that Joan of Arc suffered from 'post-traumatic fragmentation' associated with post-traumatic stress disorder. This was due to child abuse, parental neglect and other terrifying experiences. During the Hundred Years War life ranged from the horrible to the terrible as death, disease, disaster and destruction were everywhere. While the common folk suffered tragically and endlessly, not knowing what the fighting was really all about and never having 'victory' clearly defined, their leaders merely repeated for more than 100 years 'Fight on, and on, and on, and on'. 'Voices' for continued war without an end in sight sound very familiar today.[29, 30]

ARE MEN OR WOMEN MORE MUSICAL?
(Asked by Nicole Lanier of Princeton, New Jersey, USA)

This depends on how the term 'musical' is defined. Certainly there are gender differences related to music. For example:

- Studies show that a higher proportion of women than men can sing in tune.
- Generally speaking, women tend to be better dancers than men. Researchers think this is because women have a better sense of balance and a lower centre of gravity (which is also why when women collapse they tend to fall backward. When men collapse, they tend to fall forward).
- Males are more likely than females to play rock music at potentially dangerous volumes. This is regardless of whether or not they even like rock music in the first place.[31]
- When hearing music while they're placed on hold on the telephone, males prefer classical music while women prefer light

jazz. When placed on hold for the same period of time, males hearing classical music judge the wait to be shorter, while females hearing classical music judge the wait to be longer. When the music is switched to light jazz, the judgment is reversed.[32]

CAN YOU JUDGE A PERSON'S PERSONALITY AND CHARACTER BY THE SHAPE OF THEIR EARS?

(Asked by Ronnie Lathrop of Evanston, Illinois, USA)

A century ago the pseudo science of phrenology was flourishing. According to phrenologists, a person's personality and character could be identified by examining the shape and contours of their skull. Phrenology has long been discredited. So the enterprise of having the same expectations from the ears, the mouth, the nose, or any other part of the face and head seems equally doomed. But this is not the unanimous opinion of all experts. When it comes to determining personality and character, the ears in particular may give some modest but detectable indication.

According to evolutionary biologist Dr John Manning,[33] tiny swellings in a man's facial features can expose 'miserable moods' or 'a tired mind'. And women could be signalling the way to the bedroom with similar, subtle changes that conspire to make them most attractive at their most fertile times. Dr Manning has blamed 'cyclical asymmetry' for this. Cyclical asymmetry involves hormonal changes that make human tissue shrink or swell slightly out of proportion, which subtly alters the way we look. As early as 1997, Dr Manning said that 'lying on top of bone is soft tissue and this is subject to changes in size because of hormonally driven water retention and loss. It is in the face, nostrils, and particularly in the ears where you see this most clearly. It only takes a shift of a millimetre or so to change the symmetry. Symmetry is known to make us more attractive to the opposite sex'. Dr Manning adds that closely matching eyes, ears or legs signal strong genes that suggest the healthy, robust and ideal partner. Dr Manning notes that women have a rush of progesterone just after they ovulate putting them at their symmetrical peak once a month. But men are ruled by a 24-hour hormonal rhythm.[34, 35, 36]

DOES EVERYONE PRODUCE THE SAME KIND AND THE SAME AMOUNT OF EARWAX?
(Asked by Jason Waghorn of Coldstream, Scotland)

With an ear to the ground, questions about earwax (cerumen) continue to come in. Earwax is sticky, ugly and yucky, but it is important to health. Earwax is sticky on purpose because dust, dirt, bacteria, fungi and other foreign dangers to the body all stick to the wax, so that they don't enter the ear, one of the most sensitive areas of the body (and quite exposed when you think about it). Earwax also contains special enzymes called lysozymes, which break down the cell walls of foreign bacteria. Lysozymes are also contained in saliva. So earwax fights bacteria in two ways: it kind of acts as flypaper to halt them and then biochemically dissolves them.

There are different colours of earwax among various peoples of the world. In white and black people, earwax is honey-coloured, moist and soft. But in certain Asian groups (Mongolians, for instance), it is grey, dry and brittle. There is a specific gene for earwax. Wet is the dominant trait, dry is recessive. Although cosmetically we are taught to clean our ears of earwax, from a health standpoint it is probably a good idea to leave some of it there.

◆

There are 2000 glands that secrete wax in each of our ears.

◆

Research indicates that more earwax is secreted when you are afraid.

◆

It has been estimated that if you yelled for 8 years, 7 months and 6 days you would have produced enough sound energy to heat one cup of coffee.[37, 38]

◆

The BadVibes project at the University of Salford in the UK is attempting to identify the world's worst sounds and discover how evolution shaped the sound likes and dislikes of the human ear. Among the contenders for worst sound are babies crying, brakes screeching, electricity mains humming, farting, fingernails across a chalkboard, microphone feedback, seesaws squeaking, soap-opera arguments, violins misplaying and vomiting.

Chapter 6
The Mouth

WHY DO HUMANS KISS?
(Asked by Shauna McInerney of Southampton, UK)

Research reveals that kissing stimulates the same part of the brain as parachuting, bungee jumping and long distance running. Kissing is pleasurable in all sorts of ways. Here are 50 things you may not know about kissing:

1. Philematology is the scientific study of kissing.
2. Philemamania is the obsession with kissing.
3. Philemaphobia is the fear of kissing.
4. People who wake up to a kiss begin the day with a more positive attitude.
5. People who kiss their partner every morning take fewer sick days from work, have fewer car accidents on their way to work and live about five years longer.
6. A 1-minute kiss burns 109 kj (26 calories).
7. An adult kiss lasts an average of 4.5 seconds.
8. Kissing and firing a gun produce the same hormone in the body.
9. When giving a passionate kiss, you use all of your facial muscles.
10. Some scientists believe that kissing improves the skin, helps circulation, reduces blood pressure, relieves stress, alleviates headache pain, and makes us feel younger.
11. It has been theorised by some scientists that our brain may be equipped with neurons that help us find the lips of our lover in the dark.
12. Fifty per cent of all people kiss before the age of 14.

13. The typical person spends 336 hours of their life kissing.
14. Some scientists argue that infants who are kissed regularly usually develop a greater capacity in life for intelligence, artistic expression and critical thinking.
15. Over 5 million bacteria pass between 'kissers' in just one passionate kiss.
16. Infectious glandular fever (mononucleosis) and herpes are diseases that can be transferred during a kiss.
17. Kissing can help fight tooth decay by stimulating the mouth to produce more saliva. Saliva helps cleanse the teeth of harmful bacteria.
18. You're more likely to catch a common cold by shaking hands than by kissing.
19. Refusing to kiss someone under the mistletoe will not bring you bad luck.
20. If a bride does not cry when the groom kisses her at the altar, the marriage will not necessarily be unhappy.
21. If your nose itches, you will not necessarily be kissed by a fool.
22. In early Rome, the wedding kiss represented a legal bond that sealed the marriage.
23. Roman emperors used to rank a person's importance by the area of the body he or she was allowed to kiss. Important people kissed the emperor's lips, the less important kissed his hands and the least important kissed his feet.
24. Kissing became an accepted sign of affection in Europe in the 6th century.
25. Egyptians in ancient times kissed with their noses as do Inuit people (Eskimos) today.
26. The Chinese didn't kiss until they saw Westerners doing it.
27. According to legend, any person who kisses the Blarney Stone in the 15th century Blarney Castle in Cork, Ireland, will be endowed with the gift of eloquence and persuasive flattery.

28. The tradition of kissing under mistletoe stems from the tradition of slaughtering an ox under mistletoe.

29. Bonobo chimpanzees kiss to reduce stress.

30. If a dolphin likes a human enough, it will kiss them.

31. Sixty-three per cent of dog owners kiss their pooch regularly.

32. A French kiss is one in which the two kissers touch tongues. But the French did not invent it.

33. Men who kiss their wives goodbye in the morning earn much more money than those who don't.

34. The *Ananga Ranga*, a manual much like the *Kama Sutra*, suggests couples kiss in the midst of an argument so they forget what they were fighting about.

35. Basketball kissing is a sport in which couples have to get as many balls through the hoops as possible while kissing. Who keeps score, you ask?

36. The world famous Hershey's Kisses have been produced by the Hershey's chocolate company since 1907. 'Kisses' was the name given by the factory workers who noticed that the machine making the confectionary looked as if it was kissing the conveyor belt.

37. Canadian porcupines kiss one another on the lips.

38. In medieval Italy if a man and a woman were seen kissing and embracing in public they could be forced to marry!

39. The first kiss ever shown on screen was in a movie called *The Kiss* in 1896.

40. The most kisses in a single movie is 127. The film was the silent classic 'Don Juan' (1926) starring the great John Barrymore in the title role.

41. Inupiak, Polynesians and Malaysians are among those people who rub noses instead of kissing.

42. Ancient Romans kissed each other on the eyes or the mouth as a greeting.

43. In pre-revolutionary Russia, the highest sign of recognition was a kiss from the Czar.

44. Victorian etiquette required a man to kiss the back of a lady's hand.

45. A standard greeting in Europe is still a kiss on both cheeks. It could be two, four, six or whatever.

46. Some African tribes pay homage to their chief by kissing the ground where he has walked.

47. The longest underwater kiss was 2 minutes and 18 seconds and took place on 2 April 1980 in Tokyo.

48. In the US state of Indiana it is illegal for a man with a moustache to 'habitually kiss human beings'.

49. In the US city of Hartford, Connecticut, it is illegal for a husband to kiss his wife on a Sunday.

50. In the US city of Cedar Rapids, Iowa, it is a crime to kiss a stranger.

IS KISSING HARMFUL?
(Asked by Shauna McInerney of Southampton, UK)

Apart from circumstances such as two teenagers with orthodontic braces kissing each other and getting stuck, kissing is safe enough for most people to gladly take the risk. However, some communicable diseases can be transmitted via kissing. The danger is chiefly from saliva that may be exchanged. For example, Drs L.E. Cuevas and C.A. Hart[1] warn that bacterial meningitis can be transmitted through kissing. They argue that an outbreak of this disease among children in UK daycare centres was at least partly attributable to the fact that young children at the centres often kissed each other.[2]

CAN YOU GET A FOOD ALLERGY BY KISSING SOMEONE?
(Asked by Alison Klein of Somerset, New Jersey, USA)

Apparently kissing can be a way of spreading a food allergy from one person to another. At least this is the message from research by Dr Rosemary Hallett and colleagues.[3] They reviewed the cases of 379 people who were allergic to nuts or seeds. Twenty reported having an allergic reaction after kissing someone. Presumably this was after that person had eaten something to which the other person was allergic. The Hallett team reports that in most cases the reaction was mild, causing just itching and swelling in the area kissed. But in four cases, the patient experienced wheezing. The most serious case was that of a 3-year-old boy whose mother kissed him on the cheek after she had merely tasted pea soup. The boy experienced such an allergic reaction that he had to be rushed to a hospital emergency ward. The Hallett team writes that 'the possibility of an allergic reaction to a kiss is probably far from the minds of most people with food allergies. Since one-third of our subjects had reactions while dating, teenagers and young adults in particular need to be informed about this mode of exposure to allergens'.[4]

CAN YOU PREVENT A FOOD ALLERGY BY KISSING SOMEONE?

Kissing can also prevent a food allergy or at least alleviate symptoms of allergies. When people are allergic to something, they produce up to 10 times more antibodies (IgE) than they would normally. Less IgE production means less allergic response with symptoms affected accordingly. In a Japanese study by Dr H. Kimata,[5] 24 subjects with mild eczema and 24 subjects with allergic rhinitis (runny nose) 'kissed with lovers or spouses freely for 30 minutes while listening to soft music'. When tested afterwards, their measured IgE was less. So it was concluded that 'kissing may alleviate allergic symptoms by a decrease in allergen-specific IgE production'.[6] Of course, it could have been the soft music.

CAN KISSING MAKE YOU LIVE LONGER?

Kissing can indeed make you live longer. Research shows that lowered stress, lowered cholesterol level and improved relationship satisfaction are all linked to greater longevity. Kissing accomplishes all three! In one study by Dr Kory Floyd and colleagues,[7] 52 healthy adult subjects were divided into two groups. Half were instructed to 'increase the frequency of romantic kissing in their relationships' and half were instructed to make no change in their kissing behaviour. The Floyd team discovered that after 6 weeks, 'relative to the control group, the experimental group [the 'kissers'] experienced improvements in perceived stress, relationship satisfaction, and total serum cholesterol'.[8]

WHAT ARE THE MOST WIDELY SPOKEN LANGUAGES IN THE WORLD?
(Asked by Charlene Dupree of Toronto, Ontario, Canada)

There are 6800 living languages spoken by individuals throughout the world. About 6000 of these languages have registered population figures, most of which are spoken by relatively small groups of people: 52 per cent are spoken by less than 10,000 people; 28 per cent are spoken by less than 1000 people; and 83 per cent are limited to a single country. About 95 per cent of the world's languages are spoken by only about 5 per cent of the world's population. About 5 per cent of the world's languages have at least 1 million speakers and account for about 95 per cent of the world's population.

People are often surprised when they find out the top 10 languages in the world. Some are even offended that French is not in the top 10 (it's ranked 11 with about 75 million speakers). People are also surprised to learn that they have probably never heard of one of the top 10 languages — Wu. The order is as follows:

1. Mandarin: 885 million speakers
2. Spanish: 332 million speakers

3.   English: 322 million speakers

4.   Bengali: 189 million speakers

5.   Hindi/Urdu: 182 million speakers

6.   Portuguese (tie): 170 million speakers

6.   Russian (tie): 170 million speakers

8.   Japanese: 125 million speakers

9.   German: 98 million speakers

10. Wu (Chinese): 77 million speakers

The list refers to the number of people who speak the language as their first language. English might be the most widely spoken language if a second language were included.[9]

WHAT IS THE WORLD'S LARGEST LANGUAGE VOCABULARY?
(Asked by Charlene Dupree of Toronto, Ontario, Canada)

It is definitely the English language. The largest English language dictionary is the *Oxford English Dictionary*; it contains 290,000 entries with some 616,500 word forms. At the time of Shakespeare the English language possessed only between 16,000 and 30,000 words. Since then technological innovations have caused the English language to grow enormously. For example, 'computerese' did not exist 50 years ago. But now we all know about 'spam', 'hard drive', 'bytes' and so on. According to Dr Michael Cole,[10] an 18-month-old child has a vocabulary of up to 50 words; a 6-year-old has between 8000 and 14,000 words; and a 12-year-old has about 80,000 words. Language experts now believe that an average adult English speaker may possess a vocabulary of about 100,000 words that they use and about 150,000 that they understand. Only 1 out of 10 English speakers has a vocabulary of 200,000 words, while only 1 out of 100 English speakers has a vocabulary of 300,000 words or more. At the other end of the scale, the smallest language ever recorded was spoken by a tiny tribe in the Philippines and consisted of only 32 words.[11]

ARE THERE MORE HUMAN LANGUAGES OR MORE HUMAN RELIGIONS IN THE WORLD?

(Asked by Charlene Dupree of Toronto, Ontario, Canada)

There are more human languages, with some 6800 living languages being spoken in the world. There are only 4300 practised religions worldwide.

WHAT ARE THE MOST WIDELY PRACTISED RELIGIONS OF THE WORLD?

(Asked by Charlene Dupree of Toronto, Ontario, Canada)

According to the website of Adherents, an independent, non-religious affiliated organisation that monitors the number and size of the world's religions, there are 4300 religions throughout the world. Side-stepping the issue of what constitutes a religion, Adherents divides religions into churches, denominations, congregations, religious bodies, faith groups, tribes, cultures and movements. All are of varying size and influence. Nearly 75 per cent of the world's population practises 1 of the 5 most influential religions of the world: Buddhism, Christianity, Hinduism, Islam and Judaism.

Christianity and Islam are the two most widely spread religions across the world, together covering the religious affiliation of more than half of the world's population. If all non-religious people formed a single religion, it would be the world's third largest. One of the most widely held myths among those in English-speaking countries is that Islamic believers are Arabs. In fact, most Islamic people do not live in the Arabic nations of the Middle East. The world's 20 largest religions and their believers are:

1. Christianity (2.1 billion)
2. Islam (1.3 billion)
3. Non-religious (secular/agnostic/atheist) (1.1 billion)
4. Hinduism (900 million)

5. Chinese traditional religion (394 million)

6. Buddhism (376 million)

7. Primal-indigenous (300 million)

8. African traditional and Diasporic (100 million)

9. Sikhism (23 million)

10. Juche (19 million)

11. Spiritism (15 million)

12. Judaism (14 million)

13. Bahai (7 million)

14. Jainism (4.2 million)

15. Shinto (4 million)

16. Cao Dai (4 million)

17. Zoroastrianism (2.6 million)

18. Tenrikyo (2 million)

19. Neo-Paganism (1 million)

20. Unitarian-Universalism (800,000)

WHAT ARE THOSE FLAPS OF SKIN ARRANGED IN ROWS UNDER THE TONGUE AND WHAT DO THEY DO?

(Asked by Pat Wright of New York, USA)

The area of the body underneath the tongue is a little different in everyone. Observe this area under your own tongue and compare it to others (though make sure you know the person pretty well before asking if you can peer into their mouth). The band of tissue which attaches the bottom of the tongue to the floor of the mouth is the frenulum. There are two flaps of skin underneath the tongue on either side of the bottom of the frenulum in the middle of the tongue (the submandibular duct). These two flaps of skin are the sublingual folds. They serve as part of the mucosal lining of the mouth and are meant to cover and protect the salivary glands. There are also flaps of skin on the underside of the tongue called the plica fimbriata. Underneath these flaps runs the deep lingual vein. The flaps cover and protect the vein, particularly when the tongue moves. For too many of us we move our tongue a lot — and too often.

WHY DO WE SAY 'UM', 'ER', OR 'AH' WHEN WE HESITATE IN SPEAKING?
(Asked by Tom Lanier of Austin, Texas, USA)

Not everyone says 'um', 'er' or 'ah' when they hesitate while speaking. It depends on the language. For example, speakers of Mandarin often say 'zhege' which roughly translates as 'this'. In English we say 'um', 'er', 'ah' or other vocalisations for reasons linguists are not entirely sure about. 'Um', 'er' and 'ah' contain neutral vowel sounds which are the easiest sounds to make. It may be that they can be said without a great deal of thought too. So that could be part of the answer.

'Um', 'er' and 'ah' are what linguists call 'fillers' that help conversations continue smoothly. Although we may not consciously realise it, in a two-person conversation, people speak by taking turns. When someone thinks it is their turn to talk, they do. Otherwise, they listen. A two-way conversation becomes like a tennis match. Inevitably there are short periods of silence as one person pauses to let the other one speak. But sometimes a speaker doesn't want to give up their turn and instead wants a little extra time to think about what they're going to say next. They use a filler to signal this. When a listener hears the filler, they continue listening rather than start talking.

'Um', 'er' and 'ah' are examples of phonemes. In linguistics, phonemes are the smallest meaningless speech sounds humans make. The smallest meaningful speech sounds humans make are called morphemes. Everything we humans say is either meaningless or meaningful. A lot of people never learn the difference.[12]

WHY DO WE WHISTLE?

Whistling is the uttering of a clear sound by blowing or drawing air through puckered lips. 'Whistling' is from the Old English 'hwistlian' and the proto-German 'khwis' and refers to the imitative hissing sound of a serpent. There are five types of whistling: finger (wolf), hand, palate (roof), pucker and throat. Humans whistle for many reasons, among which are to express happiness, cope with boredom, alleviate anxiety or simply enhance pleasure. Whistling is one of the earliest forms of human communication and evolved from the calls of other animals and the utterances of our primate ancestors. Entire human languages are based around whistling (whistle speech). One of the most famous of these is Silbo Gomero — the so-called 'whistling language' of the Canary Islands. It has 4 vowels, 4 consonants, and over 4000 words — all of which are whistled. Other whistling languages are spoken in Burma, China, Greece, Mexico, Nepal, Papua New Guinea and Turkey.

WHY DON'T WE WHISTLE AS MUCH AS WE USED TO?
(Asked by Nicole Haack, 5AA Radio, Adelaide, South Australia)

No scientific studies exist to back up the claim that humans whistle less, more or just about the same today compared to a generation ago. However, the popular perception certainly exists that humans now whistle less. If so, is it because people are less happy, less bored, less anxious or less desiring of pleasure? It is difficult to prove any of these. Perhaps it is partly due to technology. Radios and music players are now very portable. People used to whistle when they were walking, now they are hooked up to their iPods where they only listen. Maybe it's partly due to the fact that most popular songs today are less 'whistleable' than tunes of yesteryear. Try whistling a rap song! It may even be that we have less time for whistling today as so many of us lead increasingly busy lives. We also live more crowded lives, so perhaps we're concerned about offending people who may regard our whistling as an infringement on their personal space and privacy. Or quite simply, maybe whistling is regarded as unfashionable in today's culture. In an interview, a famous supermodel said that models never smile on the catwalk because smiling is considered by the industry to be unfashionable and unsophisticated. If it can happen to smiling, it can happen to whistling.

◆

Abnormal whistling can be a symptom of epilepsy.[13]

◆

Dr Samuel Johnson (1709–1784), the great English literary figure, suffered from involuntary whistling. It is now believe he suffered from Tourette's syndrome.[14]

◆

Whistling Face Syndrome (aka Freeman-Sheldon Syndrome) is a rare inherited disorder involving craniofacial and often limb abnormalities. The mouth is tiny and in a permanent puckering position resembling someone trying to whistle, hence the name.

◆

Russian folklore holds that whistling brings bad luck.[15, 16]

WHAT ARE TAG QUESTIONS?

We know that 'um', 'er' and 'ah' are called fillers and are used in conversations to keep the listener listening while the speaker searches for the next meaningful thing to say. Although 'um', 'er' and 'ah' are thought of as meaningless sounds (phonemes), meaningful sounds (morphemes), words and entire phrases can function as fillers. Some examples are 'right', 'sure', 'you know' and 'I mean'.

Tag questions are fillers in the form of a single word or a phrase asking for agreement. Perhaps the most notable example of this type of filler is the English tag question 'innit?'. In actuality, this is a contraction of filler 'isn't it?'. Many young people in the UK, particularly in London, are known to end nearly every sentence with 'innit?'. And the trend seems to be to drop altogether the intonation that indicates a sentence that asks a question (interrogative) and make it into a sentence that sets forth or provides an explanation (declarative). Human language is continually changing, innit?[17]

HOW DO I TASTE THINGS?
(Asked by Xander Winson of Derby, Derbyshire, UK)

This is the work of your gustatory taste receptor cells (TRCs). Taste buds nestle inside the tiny bumps (papillae) you feel on your tongue. Each taste bud is a tiny sensory organ. Located within your taste buds are TRCs; they detect sweet, sour, salty and bitter. There is also a less well known and even less understood taste that TRCs detect called umami. Umami occurs in response to monosodium glutamate (MSG). Each TRC specialises in a certain taste; some sense sweet, others sense sour and so on. Whatever is sensed is signalled to the brain by way of two nerves: one is part of the fifth facial nerve called the chorda tympani, and the other is a nerve also used in swallowing called the glossopharyngeal (rolls right off the tongue, doesn't it?). When the brain receives a signal from the TRCs, we judge something to be sweet, sour, salty, bitter, or perhaps slightly differently, umami.

WHY CAN'T I TASTE ANYTHING VERY WELL WHEN MY NOSE IS BLOCKED?
(Asked by Xander Winson of Derby, Derbyshire, UK)

This reader goes on to add 'I'm eating really rich chocolate but can barely taste a thing!'

Some of the sensations commonly assigned to the sense of taste are in reality examples of the sense of smell. Many spices have relatively little taste, but affect the sense of smell powerfully. When your sense of smell is hampered by a head cold you can't taste food as well either.

IF MY MOUTH IS DRY, WHY CAN'T I TASTE VERY WELL?

A substance can only be tasted if dissolved in water or saliva. If the mouth is dry, you can't taste very well at all.[18]

◆

Over the course of a lifetime, the average human produces between 27,000 and 40,000 litres of saliva.

WHY DO PEOPLE LOSE THEIR SENSE OF TASTE AS THEY AGE?
(Asked by Amy Charlesworth of Las Vegas, Nevada, USA)

People can lose their sense of taste as they age due to the accumulated loss of sensory cells in the nose. This loss may be as much as two-thirds of the original population of 10 million cells. Although there are many exceptions, the elderly in general are less sensitive than young people to the overall perception of the food they eat. However, some 90-year-olds may be more sensitive to smells than some 20-year-olds.

When people eat, they tend to confuse or combine information from the tongue and mouth (the sense of taste, which uses three nerves to send information to the brain) with that of the nose (the sense of smell, which utilises a different nerve mechanism). It is easy to demonstrate this confusion. Cut up an unripe pear and an unripe apple into equal bite-sized pieces. Mix the pieces together. Close your eyes. Pinch your nose closed with one hand, then use the other hand to eat the pieces of fruit, one at a time. You probably will not be able to tell the difference between the pear and the apple, but you will most likely experience the sweetness of both. Now let go of your nose but keep your eyes closed. The flavour now reveals itself. This phenomenon occurs because smell provides most of the information about flavour. Chemicals from the pear and the apple, called odorants, are inhaled through the mouth and exhaled through the nose, where they interact with special receptor cells that transmit information about smell. These odorants then interact with the receptor cells and initiate a series of events that are interpreted by the brain as a smell. Estimates for the number of odorant molecules vary, but there are probably tens of thousands of these.

In contrast, taste is limited to sweet, sour, bitter, salty and the more recently discovered umami (the taste of monosodium glutamate). The sense of smell diminishes with advancing age much more so than the sense of taste. Dr T. Manrique and colleagues[19] have explored the brain's role in this process. They conclude that it is very complicated and mysterious still. In one of their studies they write 'a peculiar organisation of the memory systems during aging that cannot be explained by a general cognitive decline or exclusively by the decay of the hippocampal function'.[20, 21]

◆

The life span of a human taste bud is 7 to 10 days. Fifty per cent of taste buds are lost by age 60.

◆

Dogs, pigs and some other mammals can taste water. Humans cannot. Humans do not actually taste the water, they taste the chemicals and impurities in the water.

◆

The average human may have as many as 10,000 taste buds. The catfish has over 27,000 taste buds.

CAN I SWALLOW WHILE STANDING ON MY HEAD?
(Asked by Troy Landis of Milwaukee, Wisconsin, USA)

Do not try it! Do not even think about trying it! But it *is* possible to swallow while standing on your head, due to the pulling action involved in the 'swallow' itself. Swallowing is an activity we do many times a day and rarely, if ever, think about it. When most people think about swallowing, they probably believe that the food simply falls down the throat (pharynx) due to gravity — rather like rain down the guttering. This is hardly the case. Instead, the food is gradually pulled down the 25-cm passage to the stomach in 5 to 10 seconds. As a result of this pulling action, it is possible to swallow food or liquid while standing on your head. But it would not be very comfortable to do so and the danger of choking would be continuous.

Swallowing is a complex act for moving food from the mouth to the stomach. A series of closures to temporarily inhibit respiration occur while constrictive and peristaltic waves move the food and drink (bolus) in rhythmic muscular contractions down the oesophagus. Liquids make the trip to be drawn into the stomach in about 1 second; semi-liquids take about 5 seconds; and solids take about 10 seconds. The whole swallow takes 10 to 12 seconds. The first phase of swallowing is the termination of chewing the bolus. The bolus is pushed to the back of the throat by the tongue; it does this by pressing against the roof of the mouth (soft palate) forcing the bolus into the throat. Next, the bolus is pushed down into the oesophagus by rhythmical contractions of the oesophageal muscles (peristaltic waves). The sphincter muscle at the entrance to the oesophagus remains relaxed in order to open the channel of the throat. The larynx must elevate to force the epiglottis to close over the airway (trachea) preventing the bolus from entering the lungs. Peristaltic waves continue to push the bolus through the oesophagus and down into the stomach. One more muscle sphincter must relax for this to occur and also to prevent the bolus from coming up again.[22, 23]

❖

An article in the *Annals of Emergency Medicine* (August 1988) concludes that, among other things, there can be a 'termination of intractable hiccups with digital rectal massage'.[24]

❖

A foetus can hiccup.

◆

A man once had the hiccups for 69 years.

HOW CAN A PERSON SHOUT WITHOUT BEING IRRITATED BY THE VOLUME OF THEIR OWN LOUD VOICE?
(Asked by Vincent Rot of Heemstede, The Netherlands)

How can we stand our own shouting? After all, we're closer than anyone else to the source of the loud noise (unless we're shouting directly into someone's ear). So if others can't stand it, how can we? The answer is actually fairly simple. It is related to another classic question: 'Why does my voice sound differently to others than it does to me?'

When we listen to our own voice, including when we shout, we are not hearing solely with our ears. We are also internally hearing a mostly liquid transmission through a series of bodily organs. Speech begins at the larynx (voice box) from which a sound vibration emanates. Part of this vibration is conducted through the air. This part is what others hear when we speak. But another part of the vibration is directed through the various fluids and solids of our head. Our inner and middle ears are located within caverns hollowed out of bone. In fact, this is the hardest portion of the human skull. The inner ear contains fluid, the middle ear contains air, and both are constantly pressing against each other. The larynx is also encased in soft tissue full of liquid. Sound transmits differently through air than through solids and liquids. This difference accounts for nearly all of the tonal variations we hear compared to what others hear.

The volume of a voice depends on many factors. One of these is the size of the resonance chamber of the upper respiratory tract. When this is constricted by a cold our voice is not as loud and sounds differently. The skull and tissues insulate us from the volume of our own voice. Also, we project our voices away and not towards our ears.

This lessens the volume of our voice that we hear. If you hold your hand in front of your mouth, your voice sounds louder since your hand reflects back the sound.[25, 26]

WHAT CAN YOU LEARN FROM THE SOUND OF SOMEONE'S VOICE?

There is considerable evidence that the sound of a person's voice reveals a great deal about the speaker. Studies have shown that a listener who hears the voice of someone else can infer the speaker's social class, various personality traits, and attributes of their emotional and mental state relating to deception. In research with experimental subjects who listen to voice samples from speakers, subjects are capable of correctly estimating (75 per cent of the time) the height, weight and age of those speakers with the same degree of accuracy as that achieved by examining photographs of those speakers. This was the conclusion of a 2002 study by Dr Robert Krauss and colleagues.[27, 28]

Research by Dr Susan Hughes and colleagues[29] found that individuals with an attractive body were rated as having more attractive voices compared with individuals with an unattractive body. As deviation from an attractive body increases, the attractiveness of the voice decreases. The Hughes team concludes that the sound of a person's voice may serve as 'an important multi-dimensional fitness indicator'.[30] Voice may be an important factor in sexual attraction. Just as people are attracted to healthy bodies, they are attracted to healthy voices. Humans tend to desire fit mates. It is in their individual interest and in their species' evolutionary interest to do so.[31, 32]

WHAT IS SELECTIVE MUTISM AND DOES IT REALLY EXIST?
(Asked by Vikki de Melendez of Clifton Park, New York, USA)

Selective mutism (SM) is a well-established psychological disorder. It is a social anxiety condition in which a person is capable of speech only with a very few people and in very few situations. Many have thought that SM is not genuine because the sufferer is fully capable of speech. The unsympathetic often believe that the SM sufferer is faking the speechlessness. But sufferers can and do speak when the people around them, and the occasion itself, is 'safe' for them. The person they respond to may be a parent, a sibling or a partner; they may even only speak inside their own home and nowhere else.

A convenient way to think of SM is to imagine someone with a phobia of, say, lightning (astraphobia) and thunder (brontophobia). They are perfectly fine in fine weather, but come a storm and they collapse into panic. There are five factors which characterise SM:

1. Consistent failure to speak in specific social situations in which there is an expectation of speaking. This is often first detected when the sufferer is a child at school.
2. An interference with educational and occupational achievement.
3. The duration of the condition must last more than 1 month.
4. Failure to speak is unrelated to lack of knowledge of, or comfort with, the language to be spoken.
5. Another communicative disorder (such as stuttering) is not present.

The diagnosis of SM has been around for over 100 years. It was earlier known as elective mutism. SM is relatively rare with only between 1 and 7 per 1000 people suffering from the disease. Slightly more females suffer from it than males. So far, no single cause has been found to account for SM. Researchers led by Dr William Sharp[33] write that 'although well documented, SM is still not clearly understood, and debate continues regarding its classification and etiology'. They add that in the past and sometimes even today, an SM sufferer was dismissed as merely being defiant, manipulative, dominating, negative, stubborn, aggressive or a combination of these.[34] A major myth surrounding SM is that the sufferer chooses to whom they speak, where and when. Not so, say researchers — and they say it to one and all.[35]

WHY DO WE OPEN OUR MOUTHS TO YAWN PROPERLY?

(Asked by Jeff Grisham of San Antonio, Texas, USA)

The involuntary act of yawning usually includes opening the mouth very wide while slowly taking in a deep breath. This contortion of the mouth puts pressure on the salivary glands and causes the eyes to sometimes tear-up; it allows the throat to open wider and tightens the muscles around the mouth to ensure that the yawn is better accomplished.[36]

◆

Reading about yawning makes people yawn.

◆

It was once observed that 55 per cent of people yawned within 5 minutes after seeing someone else yawn.

WHAT ARE THE RIDGES ON THE ROOF OF THE MOUTH?

(Asked by Lawrence Cuthbert of Jarrow, South Tynesdale, Tyne and Wear, UK)

The ridges on the roof of your mouth are called the palatal rugae. Sounds rather like a game you would play in a casino, right? The palate is the roof of the mouth, and it is comprised of the hard palate and the soft palate. Palate is from the Latin *palatum* meaning, 'roof of the mouth'. It is not widely known, but the palate rather than the tongue, was once believed to be the centre of taste. This belief goes back in European folklore to at least the 16th century. Palate also came to mean 'sense of taste', hence the English word 'palatable'.

The hard palate consists of the fixed portion of the palate (palate proper) and two bones (maxillary and palatine). The soft palate consists of the palatine membrane and muscular curtain which separate the cavity of the mouth from the pharynx. The soft palate is sometimes called the velum. Rugae is from the Latin meaning 'transverse ridges'. It is related to our present word 'rugged'. The palatal rugae are transverse ridges of the palate and are widely present in mammals. Most of the research on palatal rugae has been conducted with rats. So the next time you are looking into a rat's mouth, check out the palatal rugae!

According to a classic 1989 study on palatal rugae by researchers led by Dr G. Hauser,[37] the 'biological significance [of palatal rugae] is little understood'. In a human embryo the palatal rugae are bigger and deeper. They become less so after birth and continue to shrink into adulthood.[38] There are two untested theories as to what is the function of the palatal rugae. First, the rugged surface of the roof of the mouth may help us chew food, compared with a smooth roof surface. Second, the rugged ridges may help insulate the mouth in some way. However, there is a lack of compelling evidence supporting either theory. This is despite more than 50 years of research across several human groups — as well as all those rats.

◆

Interestingly, some anatomists believe that the size and depth of the palatal rugae may be affected by thumb-sucking in infancy.

◆

A Mongolian girl reportedly has eaten 1.55 tons of mud between 1994 and 2007 because she likes the taste.

◆

An article in *Neuropsychobiology* (1997) concludes, among other things, that chewing gum of different flavours produces slightly different brain wave patterns.[39]

◆

The strongest muscle in the body is the tongue.

Chapter 7
The Skin

WHAT LIES WITHOUT: LIFE ON THE HUMAN BODY

The skin of the human body is alive with life. It is microscopic life of all kinds. In his classic work, *Life on Man* (1969), Theodor Rosebury estimates that there are 10 million individual bacteria living on the average square centimetre of human skin. Rosebury describes all of these robust, active, fertile microscopic creatures as like a 'teeming human population during Christmas shopping'. The population of bacteria on the 2m² of skin surface of the human body varies depending on what part of the body you examine. The most bacteria-prone parts of the body are the armpits, the anal region, the pubic region and the oily sides of the nose. For example, the armpit is home to about 203,000 bacteria per square centimetre.

Each square centimetre of human skin consists of around 4 million cells, 24 hairs, 35 oil glands, 6.1 m of blood vessels, 246 sweat glands, 7480 sensory cells, 23,622 pigment cells, more than 393 nerve endings and all of that microscopic life. Although all that microscopic life is high in numbers, it is small in size and weight. Rosebury estimates that all bacterial life on the surface of the human skin would fit into a medium-sized pea and possibly weigh about as much.

It is not just bacteria that live on the human skin. We can also become infested with a variety of creatures that set up house on our skin and dine there to their little heart's content. According to Dr Jonathan Kantor,[1] there are three types of louse (*Pediculus humanus corporis*) that can live on humans. They are head lice (*Pediculus capitis*), body lice (*Pediculus humanus*) and pubic lice (*Pthirus pubis*). There are also the follicle mite (*Demodex folliculorum*) that lives on the eyebrows, the scabies mite (*Sarcoptes scabiei*) that lives everywhere, along with several other *Trombiculae* mites (chiggers, redbugs, rougets, harvest and scrub). In addition, there are the tropical rat mite (*Ornithonyssus bacoti*), the human bot (*Dermatobia hominis*), the primary screwworm (*Cochliomyia hominivorax*), and of course ticks, fleas, bed bugs and a few others. Our skin is a veritable United Nations of tiny critters.[2, 3]

WHY DO HEALTHY NEWBORN BABIES OFTEN LOOK LIKE THEY HAVE MEASLES?
(Asked by Nikki Bertrand of Montclair, New Jersey, USA)

The pores of a newborn's skin are not always completely efficient. A baby can be born with spots covering their entire body. Often they are little red spots with yellowish-white centres, which can look like measles or a symptom of an infection, but are neither. It is called neonatal urticaria — completely normal — and quickly goes away by itself. A newborn can sometimes be covered with a whitish coating called vernix. Vernix is a Middle English word related to our modern word varnish and means 'odorous resin'. This covering is composed of dead skin cells and oil-gland secretions, and some describe it as a waxy or cheesy substance. It is secreted by the foetal sebaceous glands in utero. Vernix is composed of sebum (the oil of the skin) and cells that have sloughed off the foetal skin. It is hypothesised that vernix has antibacterial properties. Vernix dries and peels away during the first week or two after birth, leaving the baby with smooth, extraordinarily soft skin.

There are a variety of temporary skin conditions that can affect a newborn's appearance, which are usually nothing to be concerned about. Milia, for instance, are tiny white cysts just below the surface of the skin of a newborn and are usually located on the nose, forehead and cheeks. They disappear in the first few weeks after a baby's birth. Salmon patches are red or pink patches of skin found on the neck, on the eyelids or on the lower to middle forehead. These patches on the face generally disappear by themselves by the end of the first or second year. Patches on the back of the neck may persist somewhat longer. If you have the slightest worry about your baby's skin, consult your family doctor or specialist paediatrician. And sleep well, worry free.

◆

Human skin is made up of three layers. The outer layer is called the epidermis, which comes from the Greek *epi* meaning 'on' and *derma* meaning 'skin'. It is a tough protective layer that contains melanin. Melanin protects the skin from the sun and gives the skin its pigmentation. Under the epidermis is the second layer called the dermis, which contains nerve endings, sweat glands, oil glands and hair follicles. Beneath the dermis is the third layer composed of subcutaneous fat. Subcutaneous means 'under the skin'.

CAN A WOMAN'S BREASTS GROW ENORMOUSLY OVERNIGHT?
(Asked by Kumari Issar of Mumbai, India)

Gigantomastia is the condition in which a woman's breasts grow dramatically very, very quickly. It is a condition that occurs during pregnancy. In 2002, Dr N. Agarwal and colleagues[4] reported a case of gigantomastia in a 24-year-old woman. She was pregnant with her second child and had a 'massive bilateral breast enlargement' during her 19th week of pregnancy. After the baby was born, it took 6 months of medical treatment for her breasts to reduce to their normal size. It should be pointed out that gigantomastia is very rare. It occurs in only 1 in every 28,000 to 100,000 pregnancies.[5,6]

CAN SOME PEOPLE SWEAT IN COLOURS?
(Asked by Raoul Pimental of Houston, Texas, USA)

Sweat is not always clear. When some people sweat, it can actually come out red, yellow, blue or black. People who sweat colours have a rare condition called chromhidrosis. The coloured sweat is secreted by the apocrine glands, which are large, branched, specialised sweat glands that secrete into the upper portion of a hair follicle instead of directly onto the skin. The coloured sweat usually is localised in the face or underarms. It is thought that chromhidrosis is caused by substances, such as iron, somehow interfering with the proper sweat gland functioning. Dr J.M. Wu and colleagues[7] report that they successfully treated a man's facial chromhidrosis with botulinum toxin (BT).[8] BT is the same toxin that causes deadly botulism and is 'the most poisonous substance known' according to Dr Stephen Amon.[9] It is said that 1 tablespoon of BT could kill every man, woman and child on the face of the Earth. But a very tiny amount cured the sweat colour problems on the face of one man!

◆

Most people have two different types of sweat glands. They have ones that react to heat and others that are activated by stress or sexual arousal. Interestingly, there is an ethnic difference in the proportion of both. In general, Japanese people, for instance, have far fewer stress-related sweat glands than do northern Europeans.

◆

Do you itch or get a fever during love making? Some women suffer from a rare ailment called Fox-Fordyce disease. This disease causes an obstruction in their sweat glands around the hair follicle and an overgrowth (hypertrophy) of ugly pink pimples on the skin that can harden (hyperkeratosis). The result is that during sexual arousal, women suffer from extraordinary itching and a fever — when and where they don't want to. According to Dr A. Boer,[10] Fox-Fordyce disease is widely known as 'apocrine miliaria' — apocrine means 'secreting' and miliaria means 'fever'. Just like the line from that song 'Fever', 'When you kiss me, ya give me fever!', with Fox-Fordyce disease, that is literally true.[11]

◆

Sweat is actually odourless. The musky smell occurs when perspiration comes into contact with bacteria on the skin. The bacteria break it down and that by-product is what you smell.

◆

The US, Canada and Australia have a combined spending of some $5 billion a year on deodorants.

◆

A pair of human feet contains 250,000 sweat glands.

◆

Trying to control perspiration odour goes back to the ancient Egyptians who covered themselves with tree resins to mask body smells.

◆

Although people say 'I'm sweating like a pig', pigs don't sweat. The only animals besides humans that sweat are apes, horses and cows.

◆

Sweat can kill. In 1990, a man in the US state of Louisiana was killed by his own sweat when so much of it dropped into the electric drill he was using he was electrocuted.

◆

Napoleon considered body odour such a turn-on that when he was returning home from a battle, he would send ahead word for his wife not to wash. The erotic attraction to sweat and body odours is called mysophilia.

WHY DON'T HUMANS MOULT?
(Asked by Lisa Blumfield of New York, USA)

Most people think birds moult and humans don't. That's because birds have feathers and humans have hairs. But humans *do* moult. We shed hairs and skin cells, so technically that constitutes as moulting. Moulting means the periodic shedding of feathers, hairs, horns, nails, shells and skin — any outer layer. Moult is from the Latin *mutare* meaning 'to change'. Of course, how long is 'periodic'? Although figures vary slightly between blondes, brunettes and redheads, each of us carries an average of about 5 million hairs. About 100,000 to 150,000 of these hairs are located on the head. We lose an average of 50 to 100 hairs each day. If we comb and brush our hair we lose more hairs.

About 16 per cent of our body weight is skin. The body devotes between 5 and 8 per cent of its metabolism to the maintenance of skin. Skin is composed of skin cells; and when a skin cell dies, it is shed. The life span of a skin cell is roughly 35 days. At this rate, everyone gets a new skin a little more than about 10 times a year. Enjoying a normal life span to age 80, you will have acquired more than 800 skins. That's enough to make any bird envious.[12, 13]

WAS HUMAN SKIN USED IN BOOK BINDING?
(Asked by Jill Pascoe of Hobart, Tasmania)

Beware! This one is not for the squeamish. The use of human skin to bind books would disgust us today. But it was fairly widely practised up until about 200 years ago, particularly for medical books. In centuries past, doctors who wrote medical books would sometimes specify that they be bound in human skin. Some doctors even participated in the preparation of human skin for use in book binding. Dr John Hunter (1728–1793), the famous anatomist, father of British scientific surgery and the person after whom the London Museum of the Royal College of Surgeons of England is named, reputedly commissioned a textbook on dermatology to be bound in human skin. The skin used was often that of a flogged prisoner, particularly a murderer, who was later executed. In 1821, John Horwood was hanged for murder in Bristol, England, and his skeleton became a prized exhibit at the Bristol Royal Infirmary. A book containing details of his crime, trial, execution and dissection was published and retained at the Infirmary; the book was bound with Horwood's skin. The tanning of the skin was the work of Dr Richard Smith, the distinguished chief surgeon at the Infirmary for nearly 50 years.

The classic medical text, *Tables of the Skeleton and Muscles of the Human Body* by Bernhard Albinus (translated from Latin into English in 1749), not only was bound in human skin, but the original white skin was dyed black. This was intended to reflect one of the subjects within: 'On the location and cause of the colour of Ethiopians and of other peoples.' Dr Victor Cornil (1837–1908), the famous professor of pathological anatomy in the Faculty of Medicine at the University of Paris and author of *Syphilis* (1882), the definitive work on the subject at the time, possessed a piece of tattooed human skin from the time of Louis XIII. He also had his copy of *The Three Musketeers* (1844) by Alexander Dumas, set during the time of Louis XIII, bound in human skin. Fortunately, our sensitivities are certainly quite different today.[14, 15]

HAS ANYONE EVER DIED OF ACNE?
(Asked by Gerhard Strasse of Hanover, Germany)

It is possible that one famous person died of acne: John Stuart Mill. A famous political and moral philosopher, Mill is best known for *On Liberty* (1859) a landmark in democratic thinking; another of his notable works is *The Subjection of Women* (1869), a pioneering achievement championing the feminist revolution 100 years ahead of its time. Mill is regarded by some as one of the most brilliant men who ever lived, and the one who may have possessed the highest IQ in history — his IQ was estimated to have been as high as 200.

According to Dr Peter Cave,[16] an authority on Mill, the great philosopher may have died of erysipelas of the face — a skin rash![17] Erysipelas is from the Greek *erythros* meaning 'red' and *pella* meaning 'skin'. It is a skin infection that is normally fairly minor and today usually successfully treated with antibiotics, either in or out of hospital. Erysipelas is usually caused by a specific Streptococcus bacterium known as Group A Streptococcus. Perhaps Mill suffered a minor cut or abrasion that might have allowed the infection to begin. His lymphatic system would have been attacked as would his immune system generally. Mill's face would have turned orange and red and would have swelled up considerably. The rash would have been elevated somewhat from the normal smooth skin surface. He would have experienced great fatigue, nausea, fever, chills and pain, and abscesses may have formed too. His weakened immune system, failing to resist the infection sweeping through his body, would eventually have been overcome by the onslaught. Sadly, Mill would have died in a particularly unpleasant and extremely hot state. The earlier name for erysipelas and one still heard today is St Anthony's Fire. No, John Stuart Mill did not die from acne — but from something that might have *looked* like a very, very bad case of acne.[18]

WHAT IS THE PURPOSE OF TICKLING?

(Asked by Lou Portman of Los Altos, California, USA)

As every child knows, tickling is the act of touching a part of the body so as to cause involuntary laughter. The subject of tickling has intrigued philosophers since antiquity, even Plato and Aristotle speculated about tickling and its purpose. Tickle is derived from the Old English word *tinclian* meaning 'to touch lightly'. It was none other than Charles Darwin who was the first scientist to seriously analyse this most peculiar human behaviour. In *The Expression of the Emotions in Man and Animals* (1872) Darwin described in detail the involuntary spasms tickling triggers in babies, children, adults and non-human primates. He concluded that tickling was an ingredient in forming and keeping social bonds. Such bonding occurs through stimulating each other to laugh and feel merry. This is particularly true for parents and children. Darwin noted that the key to success in tickling is that 'the precise point to be tickled must not be known' to the person being tickled. Thus, it is surprise rather than tactile pressure that is a key ingredient in tickling.

Subsequent laboratory experiments have found that in people who are extremely suggestible, the threat of being tickled without laying a finger on them is enough to induce hysterics. This is as effective with adults as with children and provides a clue to the fact that tickling is not merely a physical sensation as Darwin theorised. Apart from Darwin's social bond theory for the importance of tickling, there is a simpler theory. The sensation felt when being tickled is similar to the one felt when insects crawl on the body. The tickle response may be a protective warning device against the stings and bites of harmful insects.

❖

It is unknown why certain areas of the body are more ticklish than others.

❖

Men and women are equally ticklish, but a few studies suggest that, if either, men may be slightly more ticklish than women.

❖

You cannot tickle yourself. If you try, you will not succeed since there is no surprise or lack of control in the stimulation. However a few studies dispute this too.

❖

Eighty-five per cent of adults in some way or another enjoy being tickled, tickling others or watching others being tickled.

❖

Some psychiatrists argue that extreme ticklishness is a sign of unrequited love.

❖

Tickling was used as a torture by the ancient Romans.

❖

Tickling is used in sexual fetishism where it is known as 'tickle torture'.

◆

In the US town of Norton, Virginia, it is against the law to tickle a woman, but it's legal to tickle a man.

◆

Research by Dr Sarah-Jayne Blakemore[19] found that robotic arms used to tickle people are just as effective as human arms.[20]

◆

Research by Dr D.S. Bennett[21] and colleagues has demonstrated that the tickle response is well established by 4 months of age.[22]

◆

Research by Dr M. Blagrove[23] and colleagues shows that the normal tickle response may be absent in those with schizophrenia.[24, 25]

WHY DOES THE SKIN OF THE HANDS TURN WHITE WHEN YOU HAVE BEEN WASHING DISHES?
(Asked by Erin West of Oyster Bay, New South Wales)

Human skin is quite different in thickness in different areas of the body. The skin on the palms, fingers, soles and toes is somewhat thicker than in other places. Skin colour becomes white from the opaqueness produced by the increased water content of the skin. Skin that has been submerged in water, such as your hands when washing dishes, appears wrinkled and shrivelled up. Most people believe that this appearance is due to the skin shrinking, but it is quite the opposite. The skin is actually expanding from the absorption of water.

Water is absorbed by skin immersed in water through the process of osmosis. This is because the internal body fluids of humans are more concentrated than fresh water. Skin whitening, wrinkling and shrivelling up appear because these areas of thicker skin expand as they become saturated. Skin elsewhere on the body also soaks up water after prolonged immersion. However, because this skin is thinner, there is more room for moisture, and the whitening, wrinkling and shrivelling up takes longer to appear. Interestingly, this process does not occur when the skin is immersed in ocean salt water. This is because sea water is very similar to the fluids in the human body such that osmosis does not take place.

◆

The skin of the average adult has a surface area of between 1.5 and 2 m^2 . Most of this is between 2 and 3 mm thick.

◆

The average square inch of skin holds 650 sweat glands, 20 blood vessels, 60,000 melanocytes and more than 1000 nerve endings.[26, 27]

◆

Skin that is smooth and hairless is called glabrous skin.

WHY DO YOU SOMETIMES SHIVER WHEN YOU WEE?
(Asked by John Rae of Clapham, Bedford, UK)

Surprisingly, for this commonly asked question, no research has been done specifically on this topic, so we are on our own to pontificate shamelessly without research evidence to temper our answer. Low room temperature as covered parts of the body are exposed could be an obvious cause of wee shiver. More seriously, probably the shivering is an example of the human body's autonomic nervous system (ANS) at work. As it runs on automatic — hence its name 'autonomic', which literally means 'self-controlling, working independently' — we are not conscious of the ANS. The urination reflex is relayed through the ANS. The reflex is directly related in strength to the amount of stretch of the bladder. So the degree of shivering is generally related to how full the bladder is at the time of urination.

The ANA has two divisions: the parasympathetic nervous system (PNS) and the sympathetic nervous system (SNS). The SNS tends to keep the bladder relaxed and the urethral sphincter contracted. This is why we don't have an 'accident' while we're concentrating on something else. It is true that the more 'desperate' you become in response to a bulging bladder, the more the SNS acts to keep you dry. According to Dr R. James Swanson,[28] the SNS response includes the release by the brain of the chemicals doadrenal medulla catacholamines epinephrine, norepinephrine and dopamine to bring about the necessary body reactions. When the opportunity arises to allow the parasympathetic side of the ANS to take over, the change in catacholamine production probably causes shivering. Laboratory experiments which have not been undertaken would prove this beyond doubt.[29] In any case, at the moment of urination there is a slight blood pressure rise and a momentary flushing or euphoria shortly after relaxing the urethral sphincter. Some find this feeling pleasurable. At such moments, some people say 'Ah'. This same response in its most extreme forms causes fainting. All of this is the ANS doing its job.[30]

WHY ARE SOME PEOPLE MORE ATTRACTIVE TO MOSQUITOES AND GET BITTEN MORE OFTEN?
(Asked by Sarah Charles of Salisbury, Wiltshire, UK)

Science is still working on the definitive answer to this question. Actually, mosquitoes do not bite at all, they suck the blood out of their victim. Most mosquitoes are attracted to the odour of carbon dioxide (CO_2). All humans exhale CO_2 with every breath. Yet this fact hardly helps explain why some people are 'bitten' more than others. Three early theories involving gender were proposed and then discarded.

The first theory claimed that women were more likely to be 'bitten' than men because mosquitoes were supposedly repelled by the strong odours of human sweat. Since the stereotyped view was that men are more likely to wash less often and be sweatier than women, women received greater mosquito attention. But this simply was not true. Some men are bitten more than some women (and vice versa). The second theory, a variation of this gender-based theory, was that mosquitoes prefer thin-skinned people. Women generally have thinner skin than men, so women are more likely to be targeted by mosquitoes. But this too proved to be untrue. The third theorised that women had some secret hormonal attractant that brought them to the attention of mosquitoes more than men. Even menstruation and ovulation could be factors in this. But such an attractant was never found. Gender does not now seem to be the all-important factor in mosquito 'bite' susceptibility.

One current theory of why mosquitoes 'bite' some people more than others is that diet makes a person more or less attractive. Dr Randolph Morgan[31] claims that regular intake of some foods (such as yeast), which ultimately are exuded through a person's pores, changes our smell and has proven effective in deterring mosquito 'bites'.[32] More recent thinking is that substances in perfumes, soap residues, facial make-up, deodorants and other compounds on the skin resulted in someone becoming more or less attractive to mosquitoes. It appears that there are over 400 such 'magnetic compounds', some which come from within the body and some from without.[33]

Recently, researchers have found that 'masking odours' are given off by the potential victim which prevents mosquitoes from finding them. Dr James Logan and Professor Jenny Mordue[34] found that 'unattractive' individuals give off different chemical signals compared with 'attractive' individuals. They tested the behavioural reaction of yellow fever mosquitoes to the odour of volunteers. In one experiment, the mosquitoes were placed into a y-shaped tube and given the choice of moving upwind down one of two branches. The air flowing down one branch was laced with odour from the volunteer's hands; the other was without this odour. Their results suggest that differential attractiveness is due to compounds in unattractive individuals that switch off attraction either by acting as repellents or by masking the attractant components of human odour. This theory differs from that of other research groups that have suggested that unattractive individuals lack the attractive components. The researchers are now testing these theories further, using foil sleeping bags to collect whole body odours from volunteers.[35]

◆

A mosquito can 'smell' their human blood dinner from a distance of up to 50 km.

◆

A mosquito can transmit fatal diseases, among which are malaria, dengue fever and Ross River fever. At least 1 million people die of malaria worldwide each year.

◆

Researchers have found that mosquitoes are most attracted to people who have recently eaten bananas.

◆

Mosquitoes cannot transmit HIV. The virus cannot survive in the mosquito.

◆

The female mosquito needs blood for protein to make her eggs. Since male mosquitoes do not make eggs, they need no blood and do not 'bite'.

◆

In 2001, the results of a study of mosquito bites in identical and non-identical twins concluded that 85 per cent of human mosquito 'attractiveness' is genetic in origin.[36]

◆

Mosquitoes rely on sugar as their main source of energy. Both male and female mosquitoes feed on plant nectar, fruit juices and other plant liquids. Sugar is burned as a fuel for flight and must be replenished daily. Blood is needed only for egg production and is consumed less frequently.[37]

◆

Air temperature may be a factor in mosquitoes 'biting', as sometimes mosquitoes prefer cool weather! According to Dr Leslie Saul-Gershenz,[40] the *Aedes* mosquito (one of more than 3000 mosquito species throughout the world) is attracted to humans only when the temperature is below 15°C.[41, 42, 43, 44]

◆

According to Dr Steven Schutz,[38] although it was once believed that blood types were an important factor in varying attraction rates, this theory has been discredited.[39]

WHY DO SOME OBJECTS FEEL COLDER TO THE TOUCH THAN OTHER OBJECTS WHEN THEY ARE IN THE SAME ROOM AND AT THE SAME ROOM TEMPERATURE?
(Asked by David Webb of Hayle, Cornwall, UK)

Some objects *do* feel colder to touch than others when they are in the same room and at the same room temperature. The temperature of an object is dependent on its heat capacity and the rate at which heat can be removed from its surface. The energy required to heat up a certain volume of material differs. Temperature is the average kinetic energy of atoms. The heat capacity per unit volume is dependent on the number of atoms per unit volume. Heat conduction is closely related to the atom number density of a material, and depends on the structure and the capability of the atoms to move within the material and pass on their energy to other atoms. For instance, marble is a dense solid and requires a great deal of energy to pass from your body to warm up. It also has a reasonable rate of heat conduction. So the surface remains cold for some time when in contact with your skin. A piece of cloth is not a dense solid, because most of its volume is occupied by air. So the cloth warms up very quickly to the temperature of the skin at which point it prevents further heat loss and feels warm. This is why cloth makes good clothing. In reality, cloth never reaches skin temperature since it is constantly being cooled by cooler air replacing warmer air. The main cooling process for cloth is convection, not conduction. It is the opposite for marble.

Styrofoam is an insulator and a very poor conductor of heat. When you hold a styrofoam cup, the heat flows from your hand and warms the styrofoam surface. The surface of the styrofoam cup becomes nearly as warm as your hand because the heat is not conducted away quickly, so little additional heat leaves the skin. Metal is a good conductor of heat. When you touch metal, heat is carried away from your hand into the metal and away from the metal surface and your skin surface. So metal feels cool.[45]

THE ARGYRIAN CANDIDATE — HOW THE 'TRUE BLUE' POLITICAL MAVERICK GAVE THE SENATE TO THE DONKEYS

The 2006 US Congressional elections saw a change in the political balance in Washington, the House of Representatives shifted from Republican elephant red to Democrat donkey blue by a healthy majority. The US Senate shifted the same way, but by only one seat. Six of the seven tight Senate races went to the Democrats, including Montana. In that state, Republican incumbent Senator Conrad Burns was defeated by Democrat Jon Tester. The winning margin was only 2565 votes (0.6 per cent). But there was a third name on the ballot — Stan Jones — the Libertarian Party candidate. Jones received 10,324 votes (2.6 per cent). This was far beyond Tester's winning margin. It is generally regarded by political pundits that if Jones had not run, most of his votes would have gone to Burns. Burns would have won re-election and the Senate would have numbered 50 Republicans and 50 Democrats. Vice-president Dick Cheney would have wielded the critical, deciding, tie-breaking vote, and the Senate would have stayed Republican instead of Democrat.

The role of Stan Jones in this historical event is pivotal. Jones did to Burns in Montana in 2006 what Ralph Nader did nationally to Gore in 2000 and Kerry in 2004. Jones is a 'true blue' political maverick — ironic in Montana, a state known for its cattle and cattle ranches. Maverick comes to us from Samuel A. Maverick, an American pioneer famous for *not* branding his cattle. The term maverick originally referred to any unbranded range animal, especially a motherless calf.

Now its most common usage is its political meaning. Maverick refers to an independent individual who does not go along with a party, a faction or a group. And that is certainly true of Stan Jones. Among the many, 'iconoclastic' views of Jones, he supports the death penalty, and bans on abortion, gay marriage and on the requirement that a social security number must be presented to acquire a Montana hunting or fishing licence. But what is very interesting about Stan Jones is that he also suffers from argyria.

Argyria is a strange body condition caused by the ingestion of silver. It can be acquired through breathing in silver dust, silver compounds or taking some silver-laden folk medicines. Argyria is from the Greek *argyros* meaning 'silver'. It was once common among silver miners, but is relatively rare today. The most dramatic effect of argyria is that the skin and sometimes the eyes turn blue or bluish-grey. And most weird of all, once the effect occurs, it is permanent and irreversible.[46]

In 2000, Jones acquired argyria by drinking a homemade colloidal silver solution, which he manufactured in his kitchen by electrically charging two silver wires in a glass of water. There is a view among some alternative therapies advocates that colloidal silver can boost the immune system — something Jones desired. Jones was also motivated by the fear that antibiotics would soon become unavailable worldwide due to an international conspiracy of pharmaceutical firms and by a belief that colloidal silver is also a treatment against the bio-terrorism disease of anthrax. Jones claims that his permanently blue skin is genuine and as such is certainly no political stunt. He also admits to being somewhat embarrassed by it — though not enough to refrain from running for public office or to blush pink.[47] In any case, history will note that in 2006, Stan Jones became the argyrian candidate — the 'true blue' political maverick responsible for turning the US Senate Democrat, donkey and 'true blue' too.[48]

◆

Odd-looking, blue-skinned people live in the US town of Troublesome Creek, Kentucky. For more than 160 years, some of the individuals, all believed to be related to one immigrant from France, have had blue skin caused by a defective gene. Their bodies lack an enzyme needed to convert a blue protein in the blood into the red protein haemoglobin. This gives them their bluish colour.

THE SCARAMANGA DEFECT: WHY DO SOME PEOPLE HAVE THREE NIPPLES?

(Asked by Kumari Issar of Mumbai, India)

It is not widely known, but about 1 in 50 women and 1 in 100 men have a 'third nipple'. This condition goes by several names, including: supernumerary nipple, accessory nipple, pseudomamma, polythelia or polymastia. The nipple is often not entirely developed as a fully functioning nipple, sometimes being mistaken for a mole. The third nipple usually has no clinical significance as long as nothing else is wrong. All mammals can develop an extra nipple.

Nipple development is classified into eight levels. The first level is the mole-like nipple. The eighth level is a fully developed, female, milk-bearing breast in a female and a fully developed, male, non-milk-bearing breast in a male. 'Supernumerary' and 'accessory' both mean extra. Pseudomamma refers to false or unnecessary breast tissue generally. Polythelia refers to the appearance of the additional nipple alone. Polymastia refers to the appearance of the additional nipple along with the presence of mammary glands. The third nipple is merely one of those anomalies that arise in human populations. Call it a birth defect if you will, but it is not a very important one. The third nipple usually forms along the two vertical 'milk lines' which start from the armpit, run down through the breasts, and further down to the groin in both women and men.

Dr Alan Ashworth and his team of researchers[49] discovered the gene that governs the growth factor that triggers breast and nipple growth. The team named it the Scaramanga gene after the three-nippled villain played by Christopher Lee in the James Bond thriller *The Man with the Golden Gun*. The discovery allows researchers to move one step closer to growing breast tissue in the laboratory in order to better study breast cancer and other breast tissue diseases.[50]

◆

Although a third nipple usually occurs somewhere on the chest, Dr D.M. Conde and colleagues[51] reported the first case of a nipple occurring on the bottom of the foot. A 22-year-old woman sought medical attention for a lesion in the plantar region of her left foot. Instead, what the doctors discovered was 'a well-formed nipple surrounded by areola and hair'.[52, 53]

WHAT IS ALBINISM?

Albinism is a congenital hypopigmentary disorder in which there is a problem from birth of a lack of pigment of the skin, hair and eyes. Albinism is from the Latin *albus* meaning 'white'. More technical names for albinism are hypomelanism or hypomelanosis. The 'melan' in these names indicates the problem in albinism — the body's melanocytes. These are protein-producing pigment cells located at the hair roots and in the uppermost layer of the skin (the epidermis). Melanocytes determine skin, hair and eye colour. Albinism sufferers usually have the correct number of melanocytes, but have a lack of the chemical means to trigger pigment (melanin) production in the melanocytes due to a genetically caused deficiency. Their skin and hair are white. In some people, only a small proportion of the body is white because only a small proportion lacks the proper functioning melanocytes. This condition is called 'localised albinism'. In others, the entire body surface is affected. This condition is called 'universal albinism'. 'Ocular albinism' affects the eyes, where the eyes are pink. They reflect the colour of blood because the blood of the capillaries shows through the transparent parts of the eye. In the normal eye, light can be partially shaded out by the pigment of the iris. But without this pigment to help exclude light, an albinism sufferer must squint or wear protective sunglasses.

The genetic deficiency that causes melanocytes to fail to produce melanin is theorised as resulting from an absence of the enzyme tyrosinase. Albinism is a genetic disorder, it is not infectious and cannot be caught. It occurs when a person fails to get from one of their parents the gene responsible for tyrosinase production. As a genetically recessive factor, albinism can emerge in any generation at any time. The development of melanocytes is quite complicated. According to Dr C.R. Goding,[54] 'over 130 genes implicated in the development or function of this cell type have been identified to date ...'[55]

CAN A PERSON WITH ALBINISM STILL DYE THEIR HAIR WITH COMMERCIAL HAIR COLOURINGS?
(Asked by Ann Cesare of Arlington Heights, Illinois, USA)

The hair of a person with albinism may be coloured with commercial hair colouring agents in the normal way. The colourless hair seen in someone with albinism is merely the absence of pigmentation. Commercial hair colourings restore pigmentation. However, according to Alex Grahl, owner of the famous Phyzogs boutique hair salon in Sydney and international award-winning hair colour specialist, the hair of someone with albinism is 'very fine and fragile. So much care must be taken in colouring their hair'.[56]

◆

'Universal albinism' occurs in about 1 in every 17,000 to 20,000 people worldwide. Other than visual problems with light, there is no evidence that sufferers with albinism are in any way physically weaker than non-sufferers.

◆

Albinism has been known to affect mammals, fish, birds, reptiles and amphibians.

◆

It is a myth that people with albinism are sterile. About 1 in 75 people are carriers of albinism genes.

WHY DO WE LIKE TO SCRATCH A WOUND WHEN IT'S HEALING?
(Asked by Nikki Boyle of Newtown, New South Wales)

Scratching an itch is a puzzling biological and behavioural response. It is especially odd to scratch a healing wound since logic would indicate that scratching would hinder rather than help the healing process. One theory of why we itch suggests that scratching stimulates the release of endorphins. These are naturally occurring opiates which block pain sensation. We release a flood of endorphins to block the pain of the initial injury more effectively while the scratching injures our skin just a little more. Therefore the gain from the endorphins outweighs the loss from the slight injury.[57]

WHAT ARE THOSE CHERRY SPOTS ON THE SKIN?
(Asked by Gerry Ryan, 2FM Radio 'The Gerry Ryan Show', Dublin, Ireland)

Cherry angiomas (also known as senile angiomas or Campbell de Morgan spots) are those strange red dots that appear on the body during and after middle age. They look as if the end of a blood vessel is trying to break through the surface of the skin. This in fact is not too far from what is actually happening. Cherry angiomas are little tufted proliferations of blood vessels just under the surface of the skin. So it is not just one blood vessel, it is composed of several. Cherry angiomas are round in shape, often slightly raised on the skin's surface and their chief feature is their bright cherry-red colour.

Cherry angiomas happen as a result of ageing skin and can appear anywhere on the body. They are caused by genetic factors, hence run in families. They have no clinical significance and do not indicate any disease. Cherry angiomas do not go away by themselves and may be easily removed by a doctor, but only for cosmetic reasons. According to Dr Hui-Jun Ma and colleagues:[58] 'To date, little is known about the pathogenesis [origin and development] and etiology [cause] of these lesions.' They further suggest that exposure to certain chemicals can cause cherry angiomas to erupt, and cite a case of a 27-year-old man, after being exposed to the chemotherapy nitrogen mustard, breaking out in cherry angiomas all over his body. If Dr Ma and colleagues are right, then it may not be merely ageing itself that causes these 'skin cherries' to bloom.[59]

DO WE REALLY NEED A DAILY SHOWER OR BATH TO STAY HEALTHY?
(Asked by Sarah Murdoch and Karl Stefanovic from Channel 9's *Today*, Sydney)

One of the most widely held myths of modern society is that humans need to shower or take a bath each day (or even more often than that) for good health. In modern industrial society, we shower and otherwise bathe for mostly social and aesthetic reasons rather than for those of health. The general rule of thumb is: 'If you can stand it socially, you can probably get by hygienically.' But in saying this, it must be stressed that bathing is necessary as one can get skin diseases and worse from not bathing at all. But the 'one per day' frequency of showers and baths is somewhat unnecessary. Our great grandparents, grandparents and perhaps even our parents probably showered or bathed less often than we do now. It was not so many decades ago when entire families routinely bathed in a common bathtub once a week. Families were larger then, so if you were fifth or sixth in the tub — you can imagine.

Standards in the degree of tolerance of body smells emanating from ourselves and others were different then compared to those of today. The advent of indoor plumbing boosted rates of bathing. It also boosted standards of laundry cleanliness. The one (or more) bathroom(s) and the one washing machine per home made cleanliness convenient to average people for the first time in human history. Cultural expectations have shifted, especially over the last century or so, to demand a cleaner population — ourselves and others. Yet nations and cultures today still differ on the expectations of bathing. Many factors impinge on this behaviour: the amount of readily available water (desert countries often have water restrictions); availability of bathing facilities (much of the world does not have easy access to a private bathroom); occupation (physical labour versus office work); lifestyle (the more athletic, the more showers); season of the year (more bathing in hot weather than in cold); age (teenagers bathe most frequently, the elderly least frequently); religion and other cultural beliefs.

WHO ARE THE CLEANEST PEOPLE ON EARTH?
(Asked by Sarah Murdoch and Karl Stefanovic from Channel 9's *Today*, Sydney)

This title *may* go to Australians; however research is difficult to come by on this point. According to an October 2006 survey of 400 people conducted by EnergyAustralia, every Australian takes at least one shower each day. Specifically, '62 per cent of people showered once a day, 29 per cent twice a day and 9 per cent showered three times a day'.[60] These figures imply that no Australian goes more than a day without a shower. This is arguably the highest rate of national showering anywhere in the world and qualifies Australia as the cleanest people on Earth. The survey also found that women take slightly longer showers than men (but only less than a minute more) and teenagers take longer showers than people over the age of 40.[61]

◆

The skin weighs about 3.2 kg.

◆

The attachment of human muscles to the skin is what causes dimples.

◆

It has been determined that one brow wrinkle is the result of 200,000 frowns.

◆

Hippocrates, the ancient Greek 'father of medicine', believed that a woman with a flat chest could enlarge her bust by singing loudly and often.

◆

Anne Boleyn, Henry VIII's second wife, had long been believed to have six fingers on one hand. It was also believed that she used to wear special gloves to conceal the fact. However, when her body was exhumed in the 19th century, it was found that she had the normal number of fingers, though it was discovered that she had three breasts.

◆

A callous is a toughened area of skin made thick by continuous rubbing, contact and pressure. A callum, a clavi and a corn are all a form of callous. A callous will most often form on the top of toes or fingers. But they can also form on the bottom of the foot (plantar region) or the bottom of the hand (palmar region).

◆

The scientific name for a corn is heloma (Hello Ma!). Helomas come in two forms, hard and soft. A hard corn is called a heloma durum. A soft corn is called a heloma molle (Hello My Molly!).[62,63]

Chapter 8
The Hair & Nails

IS IT TRUE THAT WE ARE NOT NAKED IN THE WOMB?

Desmond Morris, the great English biologist, wrote in *The Naked Ape* (1967) that of the 193 different species of primates, only one is naked: humans. Morris meant that all other primates are covered in body hair. Ironically, we are also the only primates that cover up our natural nakedness. Actually, in a sense Morris is wrong. We are not *entirely* naked. We do have at least some body hair although not nearly as much as our non-human primate cousins. From birth we have hair on our head, and in the form of eyebrows and eyelashes. At puberty we develop hair elsewhere on our bodies. Throughout life some of us lose hair in some places and gain it in others, while some places remain with hair. And for some of us, we grow so much hair that we could truly be called hairy — almost in the gorilla sense. What most of us do not know is that, not only are we not entirely hairless as adults, we are not even naked in the womb.

About 4 months into a normal 9-month pregnancy, when the foetus is about 135 mm long and weighs about 170 g, it grows a moustache! Fine hair forms on the upper lip. Gradually, over the next month or so, this fine hair spreads to eventually cover the entire body. The unborn baby remains completely hairy for many weeks. This soft, hairy coat is called lanugo. The word is from the Latin *lan* and means 'woolly down'. This hair suit is shed before or soon after birth. Each hair of the lanugo is shed one by one and is then swallowed by the baby. The tiny hairs join mucus, bile and other products to form a black substance called meconium. Just after birth, the meconium is excreted by the newborn in its first bowel movement. You must admit that this is a rather interesting way of ridding yourself of unwanted clothes! After we shed our lanugo, most of us are born with hair only on our head, and in the form of eyebrows and eyelashes. This fine, non-pigmented hair at infancy is called vellus. Vellus is Latin and means 'fleece'. As we mature, the body replaces vellus with coarser, pigmented hair called terminal hair, which becomes our true adult body hair. Nature leaves us *almost* naked. As our hair is not nearly enough to keep us warm in all weather, we wear artificial 'hair', i.e. clothes. So we are really the *nearly* naked ape.[1, 2]

WHY DON'T WOMEN AND BOYS HAVE BEARDS?
(Asked by Melanie James of Chicago, Illinois, USA)

A physician who specialises in diseases of the hair is called a trichologist. How much hair we have and where we have it is mostly a matter of hormones. Compared with men, women possess hormones that inhibit hair growth on more areas of the body and stimulate hair growth on fewer areas of the body. One of these areas where women's hormones inhibit hair growth is the face. But with facial hair growth, there is considerable variation within genders as there is between genders. Some women have more facial hair than some men. There are also ethnic differences in women's and men's facial hair growth. The 'bearded lady' at the circus of yesteryear merely suffered from a hormone imbalance.

As for boys, the male hormones responsible for facial hair do not emerge until puberty.[3]

WILL EATING BREAD CRUSTS MAKE YOUR HAIR GROW CURLY?
(Asked by Jill Rogers of Perth, Western Australia)

There is no medical evidence to support the myth that eating bread crusts will make your hair grow curly or change hair in any other way.

Although (nothing to do with hair), it is true that the crust may be the healthiest part of the bread. Compared to the lighter part of the bread, the darker part of bread may produce more healthy antioxidants. These can help prevent the body absorbing harmful oxidising agents in the atmosphere such as ozone.

Curly hair or straight hair depends on a person's genes, so the crust of bread has no bearing on genetic make-up. The origin of the bread crust myth is thought to have emerged at least 300 years ago in Europe. At that time survival was more precarious and starvation was a much more real possibility for everyone — people can lose their hair when very sick and starving. Healthy people were more likely to have enough to eat, including bread. It was also widely believed that healthy people had curly hair. Bread crusts and curly hair were seen as being somehow related. Furthermore, it was a practice well into the 19th century for poor people to sell their hair to wig makers. The expression, 'to sell one's hair for a crust', may have contributed to the myth that eating the crust has some effect on a person's hair.[4, 5]

WHY ISN'T PUBIC HAIR ALWAYS THE SAME COLOUR AS THE HAIR ON YOUR HEAD?
(Asked by Hannah Swain of The Hague, The Netherlands)

Just as with skin, the colour of hair is determined by the amount of melanin in the outer layer (cortex) of each hair. Melanin is a protein that has colour. Black hair has the greatest amount of melanin, while white hair has no melanin. Hair gets its colour from the two types of melanin that create the variety of hair colours we see. Eumelanin (sometimes called black/brown melanin) is the darkest melanin and the one most commonly found in humans. Phaeomelanin (sometimes called red melanin) is the lighter melanin. A person's hair colour is the ratio of eumelanin to phaeomelanin. A high amount of eumelanin with little phaeomelanin results in black or brown hair. As the ratio of eumelanin to phaeomelanin lessens, the result is red, ginger and blonde hair. This ratio varies enormously among humans. This is why everyone's hair is just a little different.

Melanin is produced by a group of specialised cells called melanocytes that are located near the hair bulb. They collect and form bundles of a pigment protein complex called melanosomes. The size, type and distribution of the melanosomes will determine the type of melanin produced and the ratio. The type of melanin of a person's hair is inherited. Melanin also varies in the hair of different parts of the body. This is why pubic hair is sometimes a slightly different colour from hair elsewhere. The absence of melanin later in life causes white hair. White hair may appear on some parts of the body before others because there is variation in this too.

◆

The average human hair is 91 per cent protein.

◆

If you are average, you lose about 200 hairs per day.

◆

The average human hair is composed of 45 per cent carbon, 27.9 per cent oxygen, 15.1 per cent nitrogen, 6.6 per cent hydrogen and 5.2 per cent sulphur.[6, 7]

◆

Hoo Sateow of Chang Mai, Thailand, has the longest hair in the world. His locks measure 5.16 m (16 feet, 11 inches). His brother Yee's hair is 4.88 m (16 feet) long.

CAN HAIR TURN WHITE OVERNIGHT AS A RESULT OF SHOCK?
(Asked by Ingrid Smith of Chatswood, New South Wales)

There is no scientific evidence that hair can turn white overnight due to some traumatic experience. However, legend has it that both Thomas More (1478–1535) and Marie Antoinette (1755–1793) suffered a sudden hair colour change to white the night before their executions.

Some maintain even today that a condition called *alopecia areata* can turn hair white overnight. But this condition refers to hair *loss* not hair *colour change*. Supposedly, a Dr Douglas Nelson of Averon-Bergelle, France, once described the case of a 45-year-old French farmer whose hair reportedly went from black to white in 14 days. It stayed that way for about 6 months. Then over a period of 4 months, it grew back to complete black as before. The farmer was in perfect health, he had no illness or shock. Unfortunately, there is no documentation or explanation either. Believe it at your own risk.

Theoretically, any sudden severe shock, accident, illness or change in metabolism could make hair change colour, but it will not be visible right away. According to Dr John O'Connor:[8] 'Once your hair has grown out of its follicle, any emotional or physical trauma will not affect it. This hair is basically dead, like your nails. Yet severe adverse events could cause new hair that grows out a few weeks later to be white.'[9, 10]

WHY IS IT GOOD TO BRUSH YOUR HAIR EVEN THOUGH THE HAIR IS DEAD?

(Asked by Trina Douglass, age 12, of Santa Monica, California, USA)

Welcome to the world of trichology! Trichology is the science of hair and its diseases. *Trikhos* is Greek for 'hair'. According to Dr John Mason,[11] writing in 2002, trichology is a specialty within dermatology. Brushing your hair gets rid of some of the spikes and scales on each hair shaft, moves natural oils across the hair and scalp, makes hair smoother and results in hair looking better aesthetically.[12]

According to Dr John O'Connor: 'If you don't brush your hair, but wash it regularly, then your hair can still be healthy. The belief that dreadlocks are filthy has little basis either. It's only true if you get dirt, fleas, nits or other pests matted in the hair and don't clean them out.'[13] Non-human primates such as baboons picked out such items from the hair and bodies of members of the troop as an important grooming behaviour facilitating social relations. Often nits, fleas and other items would be eaten. Among humans, in earlier historical times and well into the last century before widespread indoor plumbing, people did not wash their hair or their bodies frequently. Often the head was shaved and wigs were worn rather than washing.

WHY IS THE HUMAN FACE HAIRLESS?

(Asked by Tom Sherwood of Worthington, Minnesota, USA)

Human females have only eyebrows and eyelashes. Human males have eyebrows, eyelashes and facial hair only as adults and only around the jaw and mouth. Most other mammals have comparatively far hairier faces. So why are humans an exception? In fact, most other higher order primates have facial hair patterns at least a little similar to our own. For instance, take chimpanzees. Comparing the face of a chimpanzee to that of a recent US president ... Be that as it may. In both non-human higher order primates and humans, hair gathers around the face in a rather similar manner. It comes in varying degrees down the brow, around the ears, down alongside the cheeks and forms a beard — at least in half of us. The evolutionary adaptive virtue of a hairless face appears to be that it enables the clearer and easier sending and receiving of social messages through facial expressions. This was and is important for survival. As humans evolved into the complex creatures we are, with a complex social organisation to match, 'reading' faces well makes a lot of sense. We need to be able to discern more about the intentions of our fellow humans and so we gain valuable clues as to the likely behaviours of others by watching their faces. We discover which individuals to trust, fear, comfort, scold and so on. We would be at a distinct disadvantage without the ability to 'read' a face. Hairlessness makes it all so much better and easier to do so.[14]

◆

If the average male never shaved, his beard would be 4 m long when he died.

◆

Men without hair on their chests are more likely to get cirrhosis of the liver than men with chest hair.

WHAT ARE FINGERNAILS AND TOENAILS ANYWAY?
(Asked by Chantal Liebert of Oceanside, California, USA)

Human fingernails and toenails are the equivalent of the claws and hooves of other animals. They form a hard protective covering for the exposed and comparatively vulnerable fingers and toes, and provide a 'handy' tool for scraping and scratching when necessary. Both fingernails on the fingers and toenails on the toes cover about half of the back (dorsal) surfaces of the ends of the end bone (terminal phalanx) of each finger and toe. The parts of the nail include:

- **Nail plate.** Each nail is composed of a plate of hard keratin which is the same substance found in hair and is continuously produced at its root throughout life.
- **Nail folds.** Except for the free edge of the nail at its furthermost end, the nail is surrounded and overlapped by folds of skin (nail folds).
- **Free edge.** The nail separates from the underlying surface at its furthermost point to form a free edge. The extent of this nail at the free edge depends on personal preference and habits.
- **Root** (matrix). The root lies at the base, beneath the nail and the nail fold. This part of the nail is closest to the skin. It is here that keratin is produced by cell division. If the root of the nail is destroyed the nail cannot grow back.
- **Lunula.** The pale, crescent-shaped, half-moon area located at the base of the nail is where the root is visible through the nail.

- **Cuticle** (eponychium). This covers the proximal (near) end of the nail and extends over the nail plate to help protect the root from infection by invading micro-organisms.[15]

CAN FINGERNAILS AND TOENAILS TELL US ABOUT OUR HEALTH?
(Asked by Chantal Liebert of Oceanside, California, USA)

Surprisingly, fingernails and toenails can reveal quite a lot about a person's health. According to Dr Alan Greene,[16] changes in the look, shape or colour of nails can indicate diseases and disorders somewhere in the body long before other symptoms show up. For example, white fingernails and toenails can indicate anaemia or kidney problems. Pitted brown spots or splits in fingernail and toenail tips may indicate psoriasis. Nails are made of protein, keratin and (here is where you might be surprised) sulphur. Fingernails grow at about 0.05 to 1.2 mm a week. Toenails grow somewhat slower. Nail differences or abnormalities are often the outcome of nutritional deficiencies or disorders.

White nails with pink near the tops can be a sign of cirrhosis of the liver. White lines across the nail may indicate liver disease. A half white nail with dark spots on the tip could indicate a kidney disorder. Abnormally thick nails might be due to the blood not circulating properly in the vascular system. Yellow nails, which can occur many years before the disorder shows up, can mean there are problems with the liver, diabetes, respiratory disorders or problems with the lymphatic system. Dark nails that are flat or thin can be a sign of a vitamin B-12 deficiency. Sometimes the nail could even be spoon shaped. Brittle nails are a sign of iron deficiency and brittle and soft nails with a shine and with no moon may indicate an overactive thyroid. Ridges in the nail or nails that separate from the nail bed could also indicate a thyroid disorder, or an infection. A thyroid disorder may also be indicated if the nails are like a bumpy road.

Nails that are very bendable can be a sign of rheumatoid arthritis. Very deep blue nail beds can indicate pulmonary obstruction or even emphysema. Nails that crack, peel or chip easily may indicate that more protein and minerals are needed in the diet. Nails that resemble hammered brass may indicate a tendency towards partial or total hair loss. Flat nails can indicate Raynaud's disease. This disease affects the circulatory system and can involve the arms and legs, resulting in the hands and feet being continually cold. Unusually wide nails that are square can mean a hormonal disorder. Red skin at the very bottom of the nail bed can indicate a connective tissue disorder. According to Dr Robert Baran,[17] 15 to 40 per cent of all nail diseases are due to fungal infections. The medical term for such nail fungal infections is onychomycosis. If you are worried about your nails, see your family doctor.[18]

◆

Fingernails are made from the same substance as a bird's beak.

◆

In China, wealthy men used to grow their fingernails to astonishing lengths. Their nails would be so long that they were covered by a silk bag. This was done for reasons of status. A wealthy man could thus demonstrate that he had servants to do everything for him.

Chapter 9
The Skeleton, Bones & Teeth

IS IT POSSIBLE TO BOOST THE PHYSICAL DEVELOPMENT OF CHILDREN?

The current generation of children is perhaps the most physically inactive in history. Beginning in early childhood, play is essential for the development of physical health. Healthy physical activity has been shown to increase social, intellectual and emotional development. Sound levels of physical activity in young children are also linked to a greater likelihood of skill progression, perceived physical competence, cardiovascular and musculoskeletal health and weight management. US research findings lead to the following conclusions about developmentally appropriate activities for infants (0–1 year), toddlers (2–3 years) and preschool children (4–6 years).

Infants and toddlers
- There is insufficient evidence regarding structured activities for increasing physical activity.
- No television for children under the age of 2.
- Develop enjoyment of outdoor activity.
- Encourage unstructured exploration.

Preschool children
- Increased and extensive opportunities for a broad range of movement are necessary, including running, jumping, hopping, skipping, kicking and climbing.
- Skills develop when children are provided with sufficient time and a positive environment in which to explore and practice.
- Fun, playfulness, exploration with safe and proper supervision.

- Unorganised play with minimal instruction.
- Run, swim, tumble, throw, catch.
- Begin walking tolerable distances.
- Reduce passive transportation (car and stroller or pram).
- Limit television screen time to less than 2 hours per day.[1]

The US National Association for Sport and Physical Education recommends that parents plan sufficient times to frequently allow children to do their own walking. Furthermore, infants often spend far too much time in strollers and prams. This is called 'strollerisation'. The stroller and pram are popular time- and labour-saving devices that also help to secure children in busy and crowded places. However, for short journeys, strollers and prams compete with opportunities for children to use large muscle groups, to improve coordination and postural development and for expending energy.[2, 3]

WHY DO WE CLAP OUR HANDS TO SHOW APPROVAL?
(Asked by Paul King of North York, Ontario, Canada)

Clapping our hands to show pleasure is a behaviour we share with other primates. Chimpanzees have long been observed exhibiting this behaviour. Human babies clap their hands in delight almost from the time they are physically able to do so. Clapping is mentioned nine times in the Old Testament — five times recommended as something to do, four times condemned as something not to do. For example, as a good thing, there is: 'Oh clap your hands, all ye people; shout unto God with a voice of triumph.' (Psalms 47:1) For this reason, clapping is still debated today by church leaders as to whether or not it is appropriate in a religious service.

Clapping fulfils many different functions. When we clap, say at a concert, we are participating as the audience by adding our own percussive accompaniment to the song in progress, marking the end of the song, and showing the musicians and others in the audience of our approval. We socially and psychologically bond with the musicians and with others in the audience. We also reduce any psychological tension or emotion that may have built up while watching the performance. The noise we make is, of course, an important part of the act of clapping. Yet in ancient Greece, people clapped to drown out poor performers as well as to applaud good ones. In fact, clapping was so important in ancient Athens that people organised themselves into groups called *claques* to clap for someone or against someone. From *claques* we get both our words 'clap' (strike) and 'clique' (an exclusive group). In ancient Greece there were even professional clappers for hire to support one side or the other in debates, theatre competitions or other performances. Theatre competitions were often won or lost based on the level of applause — the earliest form of ratings.

WHY CAN'T YOU STOP TOE-TAPPING SOMETIMES?

Painful Legs and Moving Toes Syndrome is the uncontrollable movement of the toe or toes. If it just involves one toe, it's almost always the big toe. And it almost always involves pain in the legs, hence the name. No matter what you do, you can't stop the toe movement. A famous case was presented by the late neurologist Dr Harold Klawans. He describes a patient who sought treatment for ongoing pain and non-stop movement in her left foot. The moving toes reminded Klawans of the heads of a crowd at a tennis match. The heads turn one way, then the other, then back again. He writes: 'All her toes were moving in succession, one at a time: up, then down, then back in place as the next toe began to move.' No matter what the patient did, she could not stop her toes moving. 'Even when she stood up and put weight on her foot, and even when she walked ...'[4] The syndrome was first described in 1971 by Drs A.K. Tan and C.B. Tan.[5] They write that the syndrome can follow an injury to the spinal cord, tail bone, back or foot. But sometimes it has no apparent cause — not even a psychological one. Strange![6, 7]

◆

The technical name for the big toe is the hallux.

◆

The human big toe is shaped differently from a chimpanzee's big toe to better accommodate upright walking.

◆

The size and shape of the human big toe also makes pushing off when running much more efficient and helps with balance.

◆

The human heel bone is also dramatically different from a chimpanzee's heel bone because for upright walking and running it has to carry body weight so differently.

◆

Marilyn Monroe had 6 toes on one of her feet (polydactyly).

◆

The toes of the right foot of actor Ashton Kutcher are stuck together (syndactyly).

◆

Your big toe has 2 bones, the other 4 toes have 3.

◆

The shank is the part of the sole between the heel and the ball of the foot.

DOES SHINING A TORCH ON THE BACK OF YOUR KNEE PREVENT OR CURE JET LAG?

(Asked by Vanessa Wilkins of Greenwich, Greater London, UK)

This is an urban myth based on a misinterpretation of a 1998 study appearing in *Science* by authors Drs S.S. Campbell and P.J. Murphy.[8] The study deals with our body clock and how it changes due to time of day, time of month, time of year and so on. Chronobiology is the study of time in relation to the body. The two scientists conducted experiments where subjects were hooked up to a fibre optic pad illuminated by a halogen lamp placed behind the knee (not a torch). This was to test if light flashed on the back of the knee could influence our circadian rhythm (our daily body clock). This is a far cry from preventing jet lag. Behind the knee was chosen because that is where many blood vessels are close to the surface of the body. However, this does not rule out other body areas where many blood vessels are close to the body's surface.

Before the Campbell and Murphy study, most researchers had maintained that such a light effect, if it occurred at all, could only occur if the light was shined on the retina of the eyes. However, Drs Campbell and Murphy show that shining a light at the back of the knee might also bring about the same kind of alterations of our circadian rhythm. They write in their obligatory and hard-to-understand academic scholarship: 'These findings challenge the belief that mammals are incapable of extraretinal circadian phototransduction and have implications for the development of more effective treatments for sleep and circadian rhythm disorders.'[9] By this they suggest that from their findings problems such as jet lag might be better understood. Better treatments could flow on. But preventing or curing jet lag is still far up in the air.[10]

IS THUMB-SUCKING GENETIC AND DOES IT RUN IN FAMILIES?
(Asked by Michael Woodhams of Palmerston North, New Zealand)

Thumb-sucking is a 'childhood body-focused behaviour'. Others are nail-biting, scratching, hair-pulling, nose-picking and a few others. Babies are born with a sucking reflex. In fact, if this reflex is not present from birth, the infant cannot feed and quickly perishes. Babies have a natural urge to suck. They will suck on just about anything that touches their lips. The urge starts to decrease after the age of 6 months. About 80 per cent of humans suck their thumbs, but most stop voluntarily between ages 3 and 6 years. So far there has been no 'thumb-sucking gene' found. However, Japanese research by

Dr S. Ooki[11, 12] involving 1131 pairs of twins found that there was a strong genetic influence in finger-sucking behaviour in 66 per cent of male twins and 50 per cent of female twins, and in nail-biting behaviour in 50 per cent of both male and female twins.

The observation has long been made that thumb-sucking behaviour seems to run in families. Recently there has been some research to back this notion. In an Austrian study examining families of children with eating disorders, it was found that children from certain types of families were statistically significantly more likely to exhibit thumb-sucking behaviour. These families are what the researchers, Dr B. Mangweth and colleagues,[13] call 'body denying' families. They are families where parents are emotionally cold, disapproving and show a lack of intimacy towards their children.[14] Family factors and the experiences of children have a bearing on thumb-sucking, nail-biting and a host of other forms of 'oral habits'. Indeed, a study of abnormal oral habits (for example, biting other people) in the children of war veterans in Iran showed that when children have brutalising experiences, they behave brutally with their bodies, including their mouths. Dr S. Yassaei and colleagues[15] write that 'the prevalence rate [of abnormal oral behaviours] was highest in children, whose family members had been both crippled and freed prisoners of war, while the rate was lowest in children whose parents had been only prisoners of war without any lasting physical injury'.[16]

❖

Handedness is determined before you are born. The hand that is preferred when thumb-sucking as a foetus is almost certainly going to be the preferred hand after birth. For example, a study by Dr P.G. Hepper and colleagues[17] found that all of 60 right-thumb-sucking foetuses turned out to be right-handed in childhood.[18, 19, 20]

❖

Right-handed people live, on average, 9 years longer than left-handed people.

◆

It is estimated that 13 per cent of humans are left-handed. All polar bears are left-pawed.

◆

Finger agnosia is the inability to recognise the names of individual fingers of your own hand or another person's hand. It is most often caused by damage to the angular gyrus of the person's dominant brain hemisphere.

◆

An article in *The Lancet* (9 November 1996) says that if you prick your finger on a chicken bone it is possible the wound will smell putrid for 5 years. The case was of a 29-year-old man and was reported by his doctors.

◆

Your left hand does an average of 56 per cent of your typing.

◆

Another name for your pinky is your wanus.

WHY DO WOMEN HAVE SMALLER FEET, ESPECIALLY SINCE WOMEN WHO HAVE NORMAL OR LARGE BREASTS HAVE EXTRA FRONTAL WEIGHT?

(Asked by Nick Pettefair of Swindon, Wiltshire, UK)

On average, human females are smaller in stature, have smaller bones, skulls and joints than human males. Women also have narrower shoulders (but wider pelvises), and tinier elbows, wrists, hands, ankles and knees (giving women a greater tendency to be knock-kneed). If this weren't enough, women also have smaller brains, hearts, lungs, kidneys and less muscle mass (but a higher percentage of body fat). Women's legs constitute only 51 per cent of their overall height compared with 56 per cent for men. Consistent with all of the above, women's feet are smaller than men's. Given the other anatomical gender differences, it would be surprising if it were otherwise.

The human skeleton is balanced for maximum efficiency. It takes something major to alter the body's engineering needs. The weight from breasts, even very large breasts, is not nearly enough of a factor. All other things being equal, a woman with very large breasts is no more likely to have a problem keeping her balance than a woman of identical height with very small breasts and with identical sized feet. And the old saying: 'Women's feet are smaller to let them stand closer to the kitchen sink', just does not measure up either.[21, 22, 23]

◆

An article in the *British Journal of Dermatology* (June 1990) indicates that, among other things, people who think they have foot odour usually do and people who do not think they have foot odour usually do not.

◆

A human can stand jumping perhaps twice his or her height. An elephant is the only animal that cannot jump.[24, 25]

◆

The flea can jump 350 times its body length. If a human could do this, it would be the equivalent of a 1.83 m (6-foot) man jumping the length of about six soccer fields.

◆

In Western nations, the average man is about 11–12 cm taller than the average woman.

◆

The average South Korean 7-year-old is about 7 cm taller than the average North Korean 7-year-old.

WHY ARE SOME PEOPLE BETTER SWIMMERS?
(Asked by Carl Smiles of North York, Ontario, Canada)

The best swimmers tend to be tall and thin, they have long arms, legs, feet and, in particular, long hands, which allow them great 'water grasp' (a very small hand movement keeps them afloat). The best swimmers have great strength, endurance and insulation against heat loss while in water below body temperature, so they have better energy conservation. In addition, great swimmers have low resistance when in the water and can move their shoulder and elbow joints very well, so they have excellent stroke mechanics when they swim.

Part of the reason some people are better swimmers than others has to do with body density too. An average person's body density is slightly less than that of water. As muscle has greater density than fat, very muscular people tend to be poor at staying afloat. Bone also has greater density than fat, so very skinny people also tend to be poor at staying afloat. Good buoyancy is not necessarily the most important factor in good swimming, but it certainly helps. For example, great competitive swimmers — tall, thin and usually more muscular than average — tend to have greater body density than average and less buoyancy. Many would not float very well if they remained motionless in the water, though their much greater 'water grasp' makes up for it.

The average woman contains a higher proportion of fat in her body than the average man; while the average man contains a higher proportion of muscle in his body than the average woman. So in general, women are better floaters than men, but buoyancy is only one factor in swimming. Nevertheless, compared with other competitive sports, the performance of women is closer to that of men in competitive swimming.

◆

The density of salt water is slightly greater than that of fresh. The body has greater buoyancy in salt water. All other factors being equal, you can swim faster in salt water than in fresh water. This is why you can easily float in the Dead Sea (due to its very high salt content).

◆

A person can swim faster in calm rather than turbulent water. In competitive swimming, the inside lanes are faster due to the fact that the lanes closer to the sides of the pool tend to be more turbulent from the waves hitting the sides of the pool and 'washing' back.[26, 27]

IS THERE AN OPTIMAL STRIDE WHERE RUNNING IS MOST EFFICIENT?

(Asked by Ville Herva of Espoo, Finland)

What makes an optimal stride? It seems nobody knows for sure. According to Dr Kevin Beck:[28] 'It would be great to answer that question, but in fact, no one knows.' As a general rule, taller runners with longer legs have longer optimal strides. On average their optimal strides are 1.4 times their leg length while running 7-minute miles. However, on an individual level, height and leg length are poor determinants of optimal stride. In a study of 10 experienced runners, the subject with the shortest leg length had the longest self-selected and the longest optimal stride. The runners, on average, chose strides just 4 cm from optimal. Beck adds that 'most experienced runners select a stride length that does not differ dramatically from the ideal'. This implies that overall running experience has something to do with the capacity to self-select an optimal length of stride.

In one study, when collegiate runners were studied from the beginning to the end of their competitive careers, researchers found that running stride lengths tended to decrease from their first year to their final year. This is in line with the findings that elite runners tend to have shorter strides than experienced but less accomplished runners. But does all of this help an individual runner to optimise his or her stride? After all, some runners self-select strides that are shorter than optimal. Beck further adds that 'without true predictors of what an optimal stride length is for an individual (and even a trained coach would probably have a hard time determining whether a runner was over-striding or under-striding), there's not much a runner can do except let his or her body adjust to an optimal stride through experience'. It is possible that runners choose a stride rate that is most efficient, regardless of the speed, and adjust stride length to obtain the desired speed. According to Beck, 'this is because stride length has to balance with stride frequency, or stride rate, to produce a given speed, and because each runner's stride length varies widely across speeds while stride rate stays relatively constant — increasing slightly with increasing speeds'.[29]

IS THERE AN OPTIMAL SPEED WHERE RUNNING IS MOST EFFICIENT?

(Asked by Ville Herva of Espoo, Finland)

Humans adjust their walking and running gaits to minimise the metabolic energy cost of motion. The walking speed that we tend to prefer is the one that minimises energy cost per unit distance. According to Dr R. McNeill Alexander:[30] 'When time is valuable, faster speeds might seem preferable. At speeds up to 2 m/s, walking requires less energy than running, and we walk. At higher speeds, running is more economical, and we run. At each speed we use the stride that minimises energy costs.' From the use of a computer model that predicts metabolic rates for all conceivable gaits of a simple biped, we understand these and other features of human gait. For example, the energy cost of walking is increased on uphill slopes and also on soft ground such as sand. Energy expended is greater with an increased heavy load as well.[31]

◆

Humans may be the most efficient of all animals at one type of running: trotting. We may not be the fastest runners over a short distance, but we trot the best over long distances. Early humans often survived by wounding a large game animal during a hunt and tracking it down often over great distances until it fell. We did this thanks to our ability to trot.[32, 33]

◆

Humans have been walking as we do today for about 3.2 million years. This is according to research headed by Dr William Sellars.[34, 35, 36]

WHAT IS THE DIFFERENCE BETWEEN TEETH AND BONES?
(Asked by Len Newberry of Stockton, California, USA)

Bone is the calcified tissue of the skeleton of all animals with backbones (vertebrates) including humans. Bone is composed mostly of calcic carbonate, calcic phosphate and gelatine. A tooth is not a bone, although teeth grow in the jaw and contain tissue that resembles bone. Teeth are not considered a part of the skeletal system either. Histology is the study of cells and tissue on the microscopic level. This is where all the really important body stuff takes place. In terms of histological development, teeth actually form as an outgrowth or projection of the skin. Teeth represent in vertebrates the modified descents of skin plates similar to the scales of fish. It is not widely known but human teeth and the scales of a shark are much the same in basic structure. A tooth is more closely related to skin than it is to bone.

Teeth are composed mostly of a yellowish substance called dentin, which is bonelike, but not actually bone. For one thing, dentin is much softer than bone. Each tooth has a layer of white and very hard enamel to protect the dentin. Enamel, which is much harder than bone, surrounds each tooth. In fact, tooth enamel is the hardest substance that the human body produces. It has to be hard for the purpose of biting and grinding food over many years. It is the appearance and strength of tooth enamel that has led to the erroneous impression among many of us that teeth and bones are made of the same thing. Anthropologists love teeth since, being so hard and lasting so long, they make excellent fossils.

HOW ARE TEETH CONNECTED TO THE JAW?
(Asked by Len Newberry of Stockton, California, USA)

Teeth are not really connected to the jaw in any direct way. Instead, they are semi-permanently fixed within separate sockets which originate in epidermal layers in the jaw. Although teeth are harder than bone, the position of a tooth can be changed much more easily than the position of a bone can be changed.[37]

CAN FLOSSING YOUR TEETH PREVENT A HEART ATTACK?
(Asked by Ginger Whitlam of Adelaide, South Australia)

There is considerable research that bacteria in dental plaque can prompt blood to clot. And the lesions brought about by gum disease can provide a route for germs to enter the bloodstream. So this would lead to the conclusion that keeping your teeth plaque-free would help against heart attack — and you would do this better by flossing. For several years now a number of studies have suggested that people with mouth infections run a higher risk of heart disease. In 2002, Dr A. Bazile and colleagues[38] found precisely that in their research.[39] 'Treating gum disease cuts heart attack risk' according to researchers led by Dr Barbara Taylor.[40, 41]

The association of heart attack risk and gum disease was first suggested by research more than a decade ago by Dr Mark Herzberg.[42] Dr Herzberg mixed a common mouth bacteria, *Streptococcus sanguis*, with blood in test tubes and found that clots began to form. Taking a closer look at the germ, he discovered that it carries a protein similar to one in blood vessels that is known to be crucial to the clotting process. Clots can be quite dangerous. In later experiments, Dr Herzberg injected rabbits with the bacteria. Within minutes, the heart and breathing rates of the rabbits sped up. An electrocardiogram detected serious heart valve abnormalities — all signs of clots in arteries. This was no doubt rather worrying for Dr Herzberg (not to mention the rabbits). Applying findings about rabbits to humans is somewhat risky, and do we really need yet another incentive to brush after meals, floss every day and get regular dental checkups?[43, 44]

WILL LOLLIES REALLY ROT YOUR TEETH?

(Asked by Katy O'Brien, age 11, of San Mateo, California, USA)

Sugar and lollies can be eaten without inviting tooth decay if the sugar is removed from the tooth surface promptly by brushing. This is according to Dr Holly Muggleston.[45] All lollies except for those for diabetics contain sugar. There is a clear connection between sugar intake and dental cavities. Bacteria in your mouth, which promotes cavities, thrive on food particles that contain carbohydrates (sugar and starch). Also important is the length of time that food stays in your mouth, so brush right away. In this case, it is not an old wives' tale. Grandma was right![46, 47]

◆

An article in the *Journal of Periodontology* (February 1990) concludes that, among other things, waxed dental floss is preferred to unwaxed dental floss by dental patients by a ratio of nearly 2 to 1.

◆

It is true that a tooth can grow out of a foot.

◆

A tooth is the only part of the body that cannot repair itself.

◆

Teeth start growing 6 months before you are born.

◆

Bone is stronger per unit than the steel in skyscrapers.

◆

The thigh bone (femur) is stronger than concrete.

Chapter 10
The Heart, Blood & Lungs

WHAT DO YOU REALLY HEAR WHEN YOU HEAR A HUMAN HEARTBEAT?
(Asked by Thomas Schantz of Montpelier, Vermont, USA)

It surprises most people to learn that the characteristic 'lub-dub' sound of the human heart is not actually the sound of the heart beating. The human ear is not able to hear the muscle of the heart (the cardiac muscle) continually and rhythmically contracting — toiling on and on at 70 to 80 contractions per minute — during our entire lives. What is really heard is the sound of the valves within the heart clamping shut to prevent the backflow of blood during the contractions of the heart. What a doctor hears through the stethoscope while examining your heart are normally two distinct audible sounds referred to as the heartbeat. The first sound is a lower-pitched and slightly prolonged 'lub' which is produced by the closure of the mitral and tricuspid valves. This is followed by the second sound, a higher-pitched and slightly shorter 'dub', which is the closure of the aortic and pulmonary valves. But the contractions of the cardiac muscle that pumps the blood are nearly silent. If you could hear that, you could probably hear the sound of blood moving through your veins — not bloody likely!

WHAT SHAPE DOES THE HUMAN HEART MOST RESEMBLE?
(Asked by Thomas Schantz of Montpelier, Vermont, USA)

The traditional heart-shape on Valentine's Day cards and elsewhere is not the true shape of the human heart. It is a matter of aesthetic opinion as to what the heart actually most resembles. But according to

Dr Richard Jonas,[1] the human heart is actually shaped like an upside-down pear.

WHY CAN I FEEL THE LEFT SIDE OF MY HEART BEATING MORE THAN THE RIGHT?

(Asked by Thomas Schantz of Montpelier, Vermont, USA)

Like the human skull, the human heart is primarily a shell. It consists of four open spaces or cavities that fill with blood. The two on the top are the left and right atria; the two on the bottom are the left and right ventricles. Of all four, the left ventricle contracts the most forcefully. That is why you feel your heart pumping more on your left side than your right.[2]

IS THE HEART LOCATED ON THE LEFT OR ON THE RIGHT SIDE OF MY BODY?

(Asked by Jackie Bainbridge of Brooklyn, New York, USA)

People are often surprised to learn that the heart is *not* located on the left side of the body. This is really odd as most people think that because the doctor places the stethoscope slightly to the left side of their chest in order to listen to their heart, the heart must be located on the left. But it is not. Imagine drawing a line down the centre of your body from head to toe. Most of the heart is actually located on the right of the line! Only the top (apex) of the heart is on the left.

The heart is not positioned straight up and down, rather it is in a somewhat slanted position. This is so that the left ventricle can push slightly upwards towards the surface making the heart's pumping job slightly easier. But this location makes it feel as if the heart is on the left. As the most powerful pumping chamber of the heart, the left ventricle makes a loud sound too, this also creates the wrong impression that the entire heart must be on the left.[3]

HOW BIG IS THE HEART?
(Asked by Jackie Bainbridge of Brooklyn, New York, USA)

The heart is about the size of your fist. Turn your open palm towards you and make a fist. That is how big your heart is. But as we all know, the heart is as big as the goodness inside it.

HOW MANY TIMES DOES THE HEART BEAT EACH DAY?
(Asked by Jackie Bainbridge of Brooklyn, New York, USA)

It is quite amazing when you realise that the heart beats about 103,000 times per day whether or not you are awake or asleep. What a loyal worker! That is really putting your heart into the job!

CAN EXERCISE REALLY MAKE THE HEART GROW LARGER?
(Asked by Pat Cole of North Curl Curl, New South Wales)

We have known for some years now that exercise can thicken the heart walls and hence increase the size of the heart.[4] A team of Italian doctors led by Dr A. Pelliccia[5] studied 947 highly trained elite athletes and concluded that 'in some highly trained athletes, the thickness of the left ventricular wall may increase as a consequence of exercise training'.[6] The Italian research team found that the athletes with the thickest left ventricular wall were those in water polo, rowing and cycling; and athletes with the smallest left ventricular wall were those in diving. Strangely, equestrian athletes had thicker left ventricular walls than alpine skiers.

More recently, according to Dr Alfred Goldberg:[7] 'People who engage regularly in vigorous aerobic exercise undergo some remarkable adaptations. Not only will they develop more mitochondria, glucose transporters and oxidative enzymes in the muscles, they will grow new capillaries in the skeletal muscles, the heart and the brain. The left ventricle of the heart will grow larger and pump more effectively as total blood plasma volume increases. The number of circulating red blood cells will also rise, improving the ability to carry oxygen. Blood pressure will go down, as will the heart rate at rest.'[8]

◆

The heart was once believed to be the seat of human emotions. Our language contains many references indicating this: heartbreak, heart warming, cold-hearted, bleeding heart, heartfelt and so on. This association between the heart and emotions probably comes from the fact that when we are emotionally overcome, we feel it in the chest, rather than in, say, the feet.

◆

The ancient Egyptians believed that the heart was where a person's soul resided. In the afterlife, a person presented their heart to the god Anubis. This is the Greek name for the Egyptian god with the body of a man and the head of a jackal. Anubis weighed the heart. If the heart was light, the person was deemed good and was entitled to go further into the underworld. But if the heart was heavy, the heart and its owner were fed to the goddess Ammit. This is the Greek name for the Egyptian goddess with the body of a lioness and the head of a crocodile (*Amheh* in Egyptian).

WHAT IS ARTERIAL PLAQUE?
(Asked by Alec Burchfield of Terre Haute, Indiana, USA)

It is a widely held misconception that arterial plaque is composed of cholesterol. More specifically, that it is composed of 'bad cholesterol' known as low-density lipoproteins (LDLs). Cholesterol is only one component of arterial plaque, the other components are: fibrin, collagen, phospholipids, triglycerides, mucopolysaccharides, foreign proteins, heavy metals, muscle tissue cells, debris of muscle tissue metabolism and calcium.

WHAT CAUSES ARTERIAL PLAQUE?
(Asked by Alec Burchfield of Terre Haute, Indiana, USA)

Arteries possess an inner muscular layer. Plaque forms when this layer is damaged and causes the cells of the layer to mutate. The cells duplicate themselves at an extraordinary rate and eventually create a bulge inside the arterial wall. These bulges are really small, benign tumours called 'atheromas'. *Atheroma* means 'lump of porridge' in Greek. They can grow so large that they can cause the inner lining of the artery to rupture. When the arterial inner lining ruptures, the bloodstream lays down clotting fibres (fibrin) to patch the tear. A fibrin 'net' is established, which catches the other components, especially calcium and fats. Gradually this debris can build up and block the artery. The role of cholesterol in this is interesting. Cholesterol is a waxy substance and may serve to protect blood cells by cushioning them against the rough and irritating surface of fibrin-laden arterial inner-lining walls.[9]

WHAT IS HARDENING OF THE ARTERIES?
(Asked by Alec Burchfield of Terre Haute, Indiana, USA)

Hardening of the arteries (atherosclerosis) happens through inflammation and calcification. A more recent name for it is Peripheral Arterial Disease (PAD). PAD occurs when plaque builds up on the walls that carry blood from the heart to the head, internal organs and limbs. It most commonly affects blood flow to the legs. Such blocked blood flow can cause pain and numbness. It can also increase the chance of a person getting an infection and then being able to successfully fight it. PAD can impair walking and feeling in the legs. The skin of the legs can show signs of PAD due to lack of oxygen and nutrients getting to the legs. When PAD is severe in the legs it is called Chronic Limb Ischaemia (CLI), and in an advanced condition, gangrene can set in. PAD is the leading cause of leg amputation. PAD can affect arteries anywhere in the body. When it affects arteries of the heart it is called Coronary Artery Disease (CORAD), and can result in a heart attack. When it affects arteries to the brain it is called Carotid Artery Disease (CARAD) and can result in a stroke.[10]

IS HEART RATE CORRELATED WITH BIRTH ORDER?
(Asked by Lotti Otunnu of Lagos, Nigeria)

Those who continue to believe that heart rate is correlated with birth order may not realise that such a notion is based on only one study. The study dates back to the 1940s and involved 778 children from the town of Hagerstown, Maryland, USA. During routine public health screening, the heart rates of 400 boys and 378 girls who attended one school were measured. It was found that first-born and second-born children had statistically significantly shorter cardiac cycles, diastole and systole compared with children who were third-born, fourth-born, or born even later in families. (It is good to remember that families were often larger in those days.) Strangely too, this heart rate difference was found to be greater in boys than in girls.

The heart is continually filling with and emptying blood. The 'cardiac cycle' is the term for this process and consists of the complete cardiac diastole, systole and all time intervals in between. Diastole is the time at which the filling of the heart's ventricles with blood occurs. Systole is the contraction time at which the emptying of the heart's ventricles occurs. In the Hagerstown study there was no explanation for the curious finding of heart rate being associated with birth order. However, speculation included a possible heretofore unknown mysterious maternal factor appearing in the second, third or subsequent pregnancies having a bearing on the development of the child's heart.[11] Despite attempts to find corroboration for the Hagerstown finding, none has been found so far — at Hagerstown or anywhere else.[12] For example, birth order was found to have no relationship to blood pressure in a 1980 study undertaken by Dr M. Higgins and colleagues from the US National Heart, Lung and Blood Institute in Bethesda, Maryland.[13, 14]

◆

Dr Sally Edwards,[15] has proposed a set of gender-specific formula for predicting Maximum Heart Rate.

For males: 210 *minus* half your age *minus* 1 per cent of total body weight *plus* 4.

For females: 210 *minus* half your age *minus* 1 per cent of total body weight *plus* 0.[16]

HOW IS THE HEART DIFFERENT FROM ANY OTHER MECHANICAL PUMP?
(Asked by Andrew Lane of North York, Ontario, Canada)

It is difficult to imagine a mechanical pump anywhere near as good in operating so well over so long as the human heart. The heart is an organ that circulates blood around the body, without it the body would not be able to receive the vital oxygen and nutrients that are carried by the blood. The blood also removes waste products. If the heart stops functioning, the body is starved of oxygen and will soon die if it is not restarted. Thus, it is one of the most important organs in the body. This applies not only to humans, but to any multi-celled animal. The human heart is made up mostly of tough muscular walls called the myocardium. This is lined on the outside by a thin layer of tissue called the pericardium and on the inside by another thin layer called the endocardium.

The human heart has four chambers, divided into left and right. Each upper chamber is called an atrium (or auricle) while each lower chamber is called a ventricle. The atria receive blood entering the heart and the ventricles pump it out. Valves in the heart make sure that the blood only flows in one direction. The muscle cells of the heart are long and tough. They contract and relax in time with each other. When the heart relaxes, blood flows into the atria from the body through the arteries. This is known as the diastole. The healthy heart has a natural pacemaker called the atrioventricular or AV node. This is an area of tissue between the atria and the ventricles that conducts electrical impulses from the atria to the ventricles. These impulses first make the atria squeeze blood into the ventricles. The ventricles then contract, pushing blood into the arteries. This is known as the systole. The process repeats itself over and over again throughout our lifetimes. The heart keeps up a steady rhythm averaging 60 to 80 beats a minute in an average adult at rest. It can go slower when we sleep or faster when we are active.

❖

The ancient Egyptians rightly understood that the heart pumped blood, but believed that it did this to give orders to the body. In a sense, they were right.

❖

Science has discredited the view that one can predict the sex of the baby by its heart rate in utero. Still, many believe it is true. They say, a little faster, a girl, a little slower, a boy.

❖

A newborn baby's heart beats at about 130 beats per minute.

❖

A woman's heart beats faster than a man's.

❖

The human heart beats about 70 times per minute. This compares with the shrew's 600 times per minute, the hummingbird's 1300 times per minute, and the blue whale's only 10 times per minute.

❖

Your heart will thump approximately 42,075,900 beats per year and about 3 billion times in an average lifetime — give or take a few million.

❖

The heart pumps about 5–6 litres of blood around the body every minute while you are at rest.

◆

The two thumping sounds of the heart are caused by the shutting of valves. They are known as 'lub' and 'dub'.[17]

WHAT MAKES A WOUND STOP BLEEDING?
(Asked by Alicia Rauzok of Greensboro, North Carolina, USA)

A wound stops bleeding due to the process of clot formation called coagulation. Coagulation is from the Latin *coagulatus* meaning 'to cause to curdle'. Blood contains an enzyme called Protease 34 kD or thrombin for short. Thrombin is made in the liver, and the only time it seems to become active in the body is when there is an open wound.

Blood is circulating tissue composed of a fluid portion (plasma) with suspended formed elements (platelets, red blood cells and white blood cells). Arterial blood is the means by which oxygen and nutrients are transported to body tissues. Venus blood is the means by which carbon dioxide and metabolic by-products are transported from body tissues to exit the body. In blood plasma there is a specific protein called fibrinogen. In an open wound the enzyme thrombin unites with the fibrinogen to form needle-like crystals called fibrin. This union forms a biochemical alliance that catches blood cells called corpuscles as they try to flood out of the body through the wound. Corpuscle is from the Latin *corpusculum* meaning 'any small particle or body'. All of this chemical action creates a plug called a blood clot. After a time, moisture is squeezed out of the clot and it contracts; this process is called syneresis. It's not widely known but the same chemical process of syneresis also happens in the formation of jams and jellies.[18, 19]

WHATEVER HAPPENED TO ALL OF THOSE HAEMOPHILIACS WHO WERE INFECTED WITH HIV BEFORE BLOOD SUPPLIES WERE SAFE?

(Asked by Alicia Rauzok of Greensboro, North Carolina, USA)

According to the National Hemophilia Foundation in New York City, nearly 90 per cent of Americans with severe haemophilia became infected with AIDS in the 1980s when the blood supply needed for transfusions was contaminated.[20] Sadly, more than 50 per cent of those infected with HIV have died. Yet there are several myths about haemophilia. In the developed world, haemophilia is not the death sentence many people believe it is. The true nature of the disorder contrasts markedly with the popular image of the disease largely derived from its historical association with Europe's royal families, especially the Russian royal family the Romanovs.

Haemophilia is from the Greek words *haima* (blood) and *philia* (to love). The disease involves a deficiency of a specific clotting factor in the blood, which only impairs clotting, but not actually a complete inability of the blood to clot. Haemophilia is a genetic-based inherited disorder that occurs mostly in males, while the defective gene is carried by females. However, women carriers can experience mild symptoms of the disease. There are two forms of haemophilia. Haemophilia A (factor VIII deficiency) occurs in about 1 in 5000 live male births. Haemophilia B (factor IX deficiency) occurs in about 1 in 10,000 live male births. The symptoms range from mild to severe. However, most haemophiliacs can now lead fairly normal lives, while some are seriously debilitated, and a few may die prematurely.

In haemophilia, coagulation time is relatively normal and bleeding is characteristically a delayed or prolonged oozing or trickling occurring after minor trauma or surgery such as tooth extraction. Even in severe haemophilia, coagulation time ranges from 30 minutes to several hours. Rarely does a haemophiliac have massive bleeding, and never does one bleed to death from a small cut as it is commonly believed. Joint haemorrhages, gastrointestinal bleeding and bruising are actually more serious concerns for the haemophiliac than bleeding to death. Modern treatment methods have significantly reduced the risks through the use of new clotting serums given during periodic blood transfusions. However, it is estimated that 70 per cent of haemophiliacs around the world do not have access to treatment that would make this disease far less fatal.[21, 22]

WHAT IS DEEP VEIN THROMBOSIS?
(Asked by David White of Ramsgate, New South Wales)

Deep vein thrombosis (DVT), or deep venous thrombosis, is the formation of a blood clot (thrombus) in a deep vein of the body. It most commonly affects the leg veins but can affect the arm or pelvis veins. DVT has become popularly known as 'economy class syndrome' because of its association with the immobility and dehydration that can occur in long distance air travel. On average, DVT occurs in about 1 in every 1000 people per year. However, according to vascular surgeons Drs K. Kasirajan and colleagues:[23] 'Nearly 1 million patients are treated for deep venous thrombosis annually in the United States.'[24] Although there may frequently be no symptoms of DVT, usually symptoms that do appear include pain, swelling, redness of the leg (or arm or pelvis), and dilatation of the surface veins. DVT can be very serious, with between 1 and 5 per cent of those with DVT dying from a pulmonary embolism — the most dangerous complication of DVT. In a pulmonary embolism, the blood clot in the leg (or arm or pelvis) becomes dislodged from its site and blocks (embolises) the arterial blood supply of one of the lungs.

Risk factors of DVT include age, immobilisation, tobacco use, female gender, use of oral contraceptives, as well as Virchow's triad — the three factors that affect blood clot formation (rate of blood flow, thickness of blood and quality of the blood vessel wall). DVT occurs far more often in older people, but in only about 1 in 100,000 of those under age 18. This is because children are relatively more active, have better quality veins, and even have a higher rate of heart beats per minute. As the population ages, more DVT would be expected to be seen, and is.[25]

WHAT CONDITIONS DISQUALIFY YOU FROM DONATING BLOOD?
(Asked by Patricia Lowe of Boston, Massachusetts, USA)

Some people are disqualified from donating blood because they have diseases that are transmissible via blood. Other potential donors are disqualified because their conditions could endanger themselves. Nations, states, provinces and local areas vary in their rules for blood donation disqualification. According to the American Red Cross:

- A person must be the minimum age of 17 in order to donate blood. There is no maximum age limit.
- A person who has AIDS or hepatitis viruses cannot be a blood donor.
- A person who has had ear, tongue, or other bodypart piercing is allowed to donate blood as long as the needle used in the piercing was sterile. If it was not, or if this is unknown, the potential donor must wait 12 months from the time of the piercing.
- An imprisoned criminal cannot be a blood donor.
- US military personnel are not allowed to donate blood for 1 year after completing service in the Middle East.
- A diabetic is allowed to donate blood. Insulin-dependent diabetics are allowed to donate blood as long as their insulin syringe, if re-used, is used only by them.
- A person who has been deferred from travel to the UK and western Europe due to concerns about Mad Cow Disease cannot be a blood donor.

- Physically small people are not acceptable as blood donors since they have lower blood volumes and may not be able to safely lose a full pint [0.47 litre] of blood.
- A person may not donate blood if they have the flu, but can donate blood *after* exposure to someone with the flu provided that the potential donor feels well and has no symptoms.
- Pregnant women and women who have recently become mothers cannot be blood donors. The safety of donating blood during and shortly after pregnancy has not been fully established. There may be medical risks to the mother and baby during this time.
- Having high or low cholesterol does not exclude a person from donating blood.
- A person who has a low level of iron (haematocrit) in their blood may be temporarily prevented from donating blood. This requirement is for the safety of the donor to ensure that their blood iron level remains within the normal range for a healthy adult.
- For almost all cancers — such as breast, brain, prostate and lung — a person may donate blood 5 years after diagnosis or date of the last surgery, last chemotherapy or last radiation treatment.
- A person is not allowed to donate blood if they have had a blood cancer, such as leukaemia or lymphoma.
- A person who has had a non-melanoma skin cancer or a localised cancer that has not spread elsewhere, may give blood if the tumour has been removed and healing is complete.
- If a person has had malaria they cannot donate blood for 12 months, because the parasite that causes malaria can lay dormant in a person's system for as long as a year.
- A person cannot donate blood while they are on antibiotics, due to the presence of the illness or infection requiring the antibiotic, as it may be transmitted through the blood.[26, 27]

◆

The storage of blood plasma in blood banks has only been possible since 1940.

◆

If laid out in a straight line, the average adult's circulatory system would be about 96,000 km. This is enough to circle the Earth two and a half times.

◆

It takes about 20 seconds for a red blood cell to circle the entire human body.

◆

Blood is 78 per cent water.

◆

Blood vessels damaged by sunburn do not return to their normal condition for up to 15 months.

WHATEVER HAPPENED TO TUBERCULOSIS?
(Asked by Angela Karam of Ramsgate, New South Wales)

Tuberculosis (TB) in the industrial developed world is no longer the scourge of humans that it once was. However, it is still in the bodies of over one-third of the world's population. According to the World Health Organization (WHO) a new case of TB occurs somewhere in the world every second of every day. In 2004, there were 14.6 million people with active TB.[28] TB stands for Tubercle Bacillus and is caused by a bacterium called *Mycobacterium tuberculosis*. Fortunately, not everyone who is infected develops the disease, or even dies if they do develop the disease and don't receive treatment.

Approximately 1 in 10 infected with TB develops the disease; this is called active TB. In such cases if the individual receives no treatment they have about a 50 per cent chance of dying. One of the most worrying problems in the international public health field today is that TB cruelly attacks people with a compromised immune system. In sub-Saharan Africa, the AIDS epidemic does precisely this, so TB often attacks weakened individuals, with fatal consequences. TB also wreaks havoc with those with substance abuse problems.

DOES TUBERCULOSIS ONLY AFFECT THE LUNGS?
(Asked by Angela Karam of Ramsgate, New South Wales)

Contrary to popular belief, TB is not a disease that only affects the lungs. Although the most common form of TB is called pulmonary TB and it does affect the lungs. Seventy-five per cent of TB cases are pulmonary. But TB can also affect the brain, the lymphatic system, the circulatory system, the genitourinary system, the bones and the joints. Pulmonary TB is the most common type of TB in the UK, Europe, Canada, the US, Australia and the rest of the industrial developed world. In nations that can afford modern systems of health care, pulmonary TB comprises about 90 per cent of active human cases.

However, the infectious bacteria can establish themselves in every human tissue or organ. For example, TB can infect the central nervous system. This discovery was made 40 years ago by Drs Dastur and Udani — they called it tuberculosis encephalopathy. Dr G.A. Lammie and colleagues[29] write that tuberculosis encephalopathy 'is now established in the tuberculosis literature'.[30] As TB is usually acquired by the inhalation or ingestion of infected material, the lungs and intestines are the most likely locations of the illness. But TB is a group of infectious diseases and it is only because of modern drugs and adequate public health care facilities that the other types of TB are so uncommon in industrial developed countries.[31]

WHAT IS THE DIFFERENCE BETWEEN TUBERCULOSIS AND CONSUMPTION?
(Asked by Angela Karam of Ramsgate, New South Wales)

TB has been known by many names, consumption is one of these. In ancient Greece TB was known as *phthisis* meaning 'consumption' and sometimes even *phthisis pulmonalis*. Among other names, TB has also been called 'wasting disease', 'white plague' and 'king's evil'. The last name also implied that it was only the touch of a king that could cure TB. Now we have fewer kings and more drugs.[32]

WHY DOES MY BREATH SEEM TO VAPORISE MORE FOR A SHORT TIME AFTER I CYCLE ON COLD MORNINGS AND A CAR PASSES ME?

(Asked by Kevin Ryan of Hervey Bay, Queensland)

You might be able to see more water droplets because of the Tyndall Effect. The Tyndall Effect is quite interesting, as we probably first notice the results of it when we are children. The Tyndall Effect is caused by reflection of light by tiny particles in suspension in a transparent medium. This effect can be easily seen as sunlight shines through a window of a dusty room. When the sunlight passes through the air it displays air-borne dust particles. The Tyndall Effect can be seen through holes in clouds or when car headlight beams are visible on a foggy night. So the light of the car passing could illuminate water vapour particles which may not be otherwise apparent. The Tyndall Effect is named after the great 19th-century Irish scientist, John Tyndall (1820–1893). Often compared to Charles Darwin and Thomas Huxley, Tyndall was perhaps the most famous science educator of the 19th century.

❖

There are 17,000 different strains of tuberculosis whose DNA has been catalogued by a US microbiologist from New Jersey.

❖

We lose half a litre of water a day by just breathing.

❖

Partly due to the fact that the heart takes up space in the chest, the left lung is a little smaller than the right lung.

◆

The total surface area of a pair of human lungs is equal to that of a tennis court.

◆

A runner consumes about 7 litres of oxygen while running a 100 m dash.

Chapter 11
The Stomach & Intestines

WHY ARE OPERA SINGERS FAT?

(Asked by Kelly Reed of Indianapolis, Indiana, USA)

There are several theories attempting to explain why opera singers are often so pleasingly plump. One holds that a large amount of fatty tissue surrounding the voice box (larynx) increases its resonance capability and thus produces a more pleasing sound. The amount of this fatty tissue varies from singer to singer. It is almost impossible to have a great deal of fatty tissue around the voice box without carrying a great deal of fatty tissue elsewhere on the body. A second theory holds that opera singers need a far more powerful diaphragm than normal to be able to project their voice above the sound of a large orchestra in a large opera house. A large chest cavity and good control of the lungs will provide a suitable mass to help drive the diaphragm to some extent. A large body mass and a large body frame to support it help even more, so there is a huge advantage in being huge. In the 18th and 19th centuries, when opera was an expanding medium, successive generations of opera producers sought larger and more dramatic effects, larger audiences and larger theatres. Being human, singers could not be re-engineered, so improved vocal techniques were developed to cope better with this steady rise in size and volume. Consistent with this, it was recognised as an advantage for a singer to have a large chest, rib cage, neck, mouth, everything. The desire for larger singers was not the only change, new technology for wind instruments and the reconstruction of older baroque string instruments also came about as the result of opera innovations.

A third theory comes from Dr Peter Osin.[1] Dr Osin argues that opera singers 'may be more predisposed to put on weight because exertions in the lungs act as a trigger for their appetite'. He adds that 'the mechanism of singing encourages the lung cells to release chemicals including leptin, a protein made by the body's fat cells that is involved in the regulation of appetite'.[2] A fourth theory holds that the act of opera singing itself expands the body, particularly the rib cage. After years of singing, the opera singer's body may look fat, perhaps fatter than it really is. This is the implication of Australian research by Dr C.W. Thorpe and colleagues.[3, 4] No theory can claim to be proven so far. Others will probably emerge. As they say: 'It ain't over until the fat lady sings.'

◆

There are exceptions to the rule that opera singers are fat. José Carreras, one of the most famous tenors of the last 50 years, is only 170 cm tall and is rather average in body mass.

◆

In 2004, opera singer Deborah Voigt was dismissed from a Royal Opera House production in London because she was too fat.

◆

Many opera devotees fondly refer to *Madame Butterfly*, one of the most famous operas and diva roles, as 'Madame Butterball'.

◆

The phrase, 'It ain't over until the fat lady sings', has been attributed to, among others, US-sports caster Dan Cook, US basketball coach Dick Motta, US baseball player Yogi Berra, US author Damon Runyon and that ageless savant: Anonymous.[5, 6, 7]

IN THE HOTTEST WEATHER, WHAT SHOULD WE DRINK TO COOL DOWN: SOMETHING COLD OR SOMETHING HOT?
(Asked by Andrew Wiseman of Cambridge, UK)

The claim is that drinking a cold drink is better since: 1) the cold drink comes in contact with the tissues of the mouth, tongue and throat thus soothing us; and 2) the cold drink brings body temperature down since the body must warm up the drink to body temperature.

On the other hand, the claim is that drinking a hot drink is better since the body has to use more energy to reduce the temperature of the drink to body temperature. But this is contradicted by those who maintain that this body process only makes you hotter. People choose with their feet on this one (well, their mouths too). People prefer cool drinks much more often than hot drinks during hot weather for sensory reasons rather than for body temperature. The same is true with hot drinks being more preferable during cold weather. In fact, the temperature of the drink does not really matter unless a massive amount of liquid is being consumed or if the temperature of the liquid is extremely cold or extremely hot. If extremely cold, such an amount would probably make you sick and cause you to vomit and cramp. If extremely hot, it would burn your mouth. The fact that fluid temperature, cooler or warmer, doesn't much matter is due to the much greater mass of the body compared to the drink. Cold or hot, the temperature of the drink is transformed to body temperature without much lowering of the body temperature at all. The body's system of temperature regulation (homeostasis) is not so easily fooled. The US Centres for Disease Control and Prevention (the CDC) stresses that during hot weather more fluids should be consumed. It makes no recommendation as to the temperature of this fluid other than to 'avoid very cold drinks', as these can cause stomach cramps. The CDC says that besides water, salts and minerals also need to be replaced in order to avoid heatstroke, so it recommends avoiding caffeine, alcohol and drinks with large amounts of sugar, because these can 'actually cause you to lose more body fluid'.[8,9]

IS IT FEED A COLD AND STARVE A FEVER OR FEED A FEVER AND STARVE A COLD?

(Asked by Elektra Filipo of Apia, Samoa)

Many people are confused about this. According to Dr Andrew

Lloyd:[10] 'There is little medical evidence to support "feed a cold, starve a fever".[11] In the context of an acute infection like influenza, the best treatment is for the patient to eat only if hungry.' According to Dr Holly Muggleston:[12] 'You do need to feed a fever. That is, you should get adequate fluid to reduce the risk of dehydration. And, in some cases, you need adequate kilojoules in order to prevent weight loss. This depends on the fever's temperature and duration.'[13] So from my humble non-physician standpoint the way to go is 'feed a fever' and 'feed *or* starve a cold'.[14]

WILL EATING SPINACH MAKE ME STRONG?
(Asked by Thomas Glass of Hartford, Connecticut, USA)

The belief that eating spinach will give you big muscles is called the 'Popeye Effect'. Unfortunately, this is a myth. Just eating spinach is not going to make anyone as strong as the cartoon character, Popeye; if it did, body builders and athletes would be popping tins of it. The fact is many other nutritional supplements build muscle bulk and strength faster and more effectively than spinach. However, it is often true that people with weak muscles have a variety of vitamin and mineral deficiencies. Spinach derives from the Latin *spina* meaning 'spine'. Spinach is a vegetarian, low-kilojoule, non-animal source of iron and magnesium, both of which are essential to muscle development. Spinach also contains vitamin C, vitamin B-9 (folate, folic acid) and additional antioxidants that the conventional wisdom in human nutrition claims helps prevent cancer. All of these are beneficial to the body. Dr Shawn Somerset[15] points out that a difficulty in over-relying on spinach as a source of iron is that spinach is rather poorly absorbed by the body unless eaten with calcium.[16] The type of iron found in spinach is non-blood (non-haeme), a plant iron, which the body does not absorb as efficiently as blood (haeme) iron, found in meat. The myth of the 'Popeye Effect' dates back to a real scientific mistake. In

1870, Dr E. von Wolf mistakenly misplaced a decimal point and wrongly estimated the iron content of spinach to be 10 times more than any other green vegetable. The mistake was corrected only in 1937, but by then it was too late. The first Popeye cartoons appeared in 1929 — and the spinach–muscles–strength legend was already born.

◆

The idea that spinach makes you strong relates to its iron content. Iron is essential for transporting oxygen from the lungs to the muscles and for storing oxygen in the muscles. The health benefits of spinach include:
* protecting against heart disease
* helping prevent colon and prostate cancer
* maintaining good eye health
* reducing chances of brain damage after a stroke
* helping preserve healthy bones.[17]

WHY DON'T WE SUFFER FROM E. COLI ALL THE TIME?
(Asked by Neil Greenwood of Cardiff, Wales, UK)

All bacteria in the gut come from the environment and are ingested during feeding. For example, while a mother's milk should be sterile, other environmental factors allow bacteria to enter. Bacteria colonise the mouth first and then are swallowed in large numbers along with the milk. E. coli (*Escherichia coli*) is one of a diverse group of micro-organisms called coliforms. Coliforms are classified mostly by the tests for their isolation rather than for any physical characteristics. They are found in the lower gastrointestinal tract of humans and animals. Coliforms are unlikely to grow in water, but they grow nicely in food especially if it has been kept at conditions that may enhance growth — room temperature on a warm day is certainly enough!

E. coli ferments and lives on the lactose of milk. Fermentation is an enzymatically controlled anaerobic breakdown or transformation of an energy-rich organic compound. Carbohydrate to carbon dioxide and alcohol to an organic acid are examples of fermentation. E. coli colonises almost immediately. Most forms of E. coli are not capable of causing disease (non-pathogenic). They are found in large numbers in the gut and can be beneficial to humans in keeping dangerous bacteria at bay and producing various vitamins that the body absorbs from the colon. However, some strains can cause disease (pathogenic), for example, E. coli 0157 and E. coli 0111. E. coli has been implicated in some severe food poisonings in every corner of the world.[18]

WHY DO YOU SOMETIMES LOSE BLADDER OR BOWEL CONTROL AT TIMES OF EXTREME FEAR?
(Asked by Peter Neaum of Albury, New South Wales)

Humans sometimes urinate or defecate at times of extreme fear due to the acute stress response. This is sometimes referred to as the 'flight or fight response' coined by Walter Cannon in 1929. When we are experiencing extreme fear, the sympathetic nervous system goes into overdrive and produces a state wherein we are better prepared to engage the source of the fear in a struggle or better prepared to flee from it. A number of temporary physiological changes occur in the acute stress response. For example, more adrenaline is pumped through the system. This boosts heart and lung activity and thus aids physical functioning needed in fighting or running. Other temporary physiological changes include dilation of blood vessels for muscles, constriction of blood vessels in parts of the body not needed for fighting or fleeing, liberation of nutrients needed for muscular action, inhibition of tear glands and salivation, dilation of the pupils, inhibition of stomach and intestinal action, inhibition of erection in males, and relaxation of elimination control. So under circumstances of acute stress, such as in extreme fear, the bladder and bowels can 'let go'.[19, 20]

◆

An article in *Military Medicine* (August 1993) indicates that constipation occurs to US military personnel at the following rates: 7.2 per cent while at home, 10.4 per cent while being transported to the field and 30.2 per cent while in the field.

◆

In an article in the *Scottish Medical Journal* (December 1993) three cases are presented of porcelain lavatory pans collapsing under body weight, producing wounds which required hospital treatment. Excessive age of the toilets was implicated as a causative factor. As many toilets get older episodes of collapse may become more common, resulting in further injuries.

◆

An article in *Tidsskift for den Norsk Laegeforening* [*The Journal of the Norwegian Medical Association*] (20 March 1999) concludes, among other things, that patients differ in which containers they prefer for urine samples. (Will the next 'must have' fashion accessory be the designer urine sample container?)

◆

It has been estimated that if you broke wind continuously for 6 years and 9 months, enough gas would be produced to create the energy of an atomic bomb about the size of those first developed.

◆

A cat's urine glows under a black light. A human's urine does not glow under any light.[21, 22]

WHAT ARE GALLSTONES?

(Asked by Sonia Axtens of Sacramento, California, USA)

Gallstones are crystalline bodies formed within the gallbladder. The gallbladder is only small, about 7.5 cm long and 2.5 cm wide, and is rather pear-shaped. The organ sits beneath the liver on the right side of the abdomen. The gallbladder stores the bile, manufactured by the liver, and then contracts and releases the bile into the intestines. This greatly assists digestion by helping to break down fats in food. The medical term for a gallstone is a choleith. In fact, 80 per cent of gallstones are cholesterol stones since they are made up mostly of cholesterol, the other 20 per cent are pigment stones as they are made up mostly of bilirubin, a reddish-yellow pigment, and calcium salts. A gallstone can be the size of a grain of sand or as large as a golf ball! They can occur one at a time or many together — even in their thousands! According to the American Liver Society, about 1 in every 12 Americans has gallstones.[23] At that rate there are 25 million people with gallstones in the US, 5 million in the UK, 2.7 million in Canada and 1.6 million in Australia.

Gallstones may form in one of three ways. When bile contains more cholesterol than it can dissolve, when there is too much of certain proteins or other substances in the bile that causes cholesterol to form hard crystals, or when the gallbladder does not contract and release bile regularly. Most of the time gallstones are painless, and this condition is called 'silent gallstones'. However, when they are painful the condition is called 'symptomatic gallstones'. Each year symptomatic gallstones trigger serious health problems, causing some 800,000 hospitalisations and resulting in more than 500,000 operations in the US alone.

ARE WOMEN WHO ARE 40, FAT AND FAIR MORE LIKELY TO GET GALLSTONES?

(Asked by Sonia Axtens of Sacramento, California, USA)

The old wives' tale that women who are 40, fat and fair are more likely to get gallstones has some basis in fact. According to Dr Terry Bolin,[24] women who are in this category are indeed at some risk of developing gallstones. However, many who are older or younger are also sometimes affected. Weight is a definite factor. Obesity, rapid weight loss, pregnancy, taking birth control pills, or undergoing hormone replacement therapy may be contributing factors to developing gallstones.[25] According to the American Liver Society, rapid weight-loss dieting (over 3 pounds [1.36 kg] per week) increases the risk of developing gallstones more so than a slower rate of weight loss.[23] However, claims Dr Bolin, fair hair or fair complexion does not increase risk of developing gallstones.[24]

◆

In folk medicines of certain cultures, bovine gallstones are believed to be an aphrodisiac. Today, the gallstones of dairy cows constitute a valuable commodity that fetch up to US$1000 per ounce. In the meat-packing industries of some nations, workers are checked before they leave to ensure they have not stolen bovine gallstones.

◆

'Porcelain gallbladder' is a condition in which deposits of calcium build up in the wall of the gallbladder causing the gallbladder to become calcified, fragile and brittle. The gallbladder becomes breakable like a small porcelain vase, hence the name. The condition was named in 1951 by Dr R.H. Kazmierski.[26] According to Dr Tsung-Chun Lee and colleagues,[27] fortunately porcelain gallbladder is rare, occurring in less than 1 in every 1000 gallbladder patients requiring hospitalisation.[28, 29]

HOW MUCH DAMAGE DOES A TAPEWORM DO TO THE HUMAN BODY?
(Asked by George Gomez of Concord, California, USA)

There are many species of tapeworms but not all can infest humans. Though surprisingly, tapeworms produce few physical problems for the human body in infected individuals. However, they are still better to be avoided. Tapeworm infestation is less of a problem in modern industrial societies nowadays since the quality of our food is higher — we can thank the inventor of the refrigerator for much of this. Tapeworms come into the body via contaminated food. Many organisms live on and in the human body so tapeworms can easily survive and thrive indefinitely inside us. A tapeworm in a human can range in length from 0.0063 cm to an incredible 15 m! Tapeworms have no digestive tract so they must eat food already digested by another animal, which is precisely what they do as a parasite inside human intestines, by absorbing nutrients directly across their skin (cuticle). They also reproduce inside the human body.

Tapeworms are simple in design, but ruthless in action. They consist of two organs. The first organ (scolex) anchors the beastie to the wall of the intestine with suckers and hooks. The second organ (proglottid) is really a series of organs that grow out from the scolex with each having full reproductive capability. Proglottids form a chain of varying lengths. The last segment of the chain eventually breaks off and is passed out with faeces. Tapeworms resist being destroyed by the body's immune system or digestive juices. Tapeworms cause health problems around the world and can even kill since they rob us of nutrients, block our intestines and take up space in organs that stop them from functioning normally. A tapeworm cyst can settle in the brain, eye, liver and elsewhere. Although there is little knowledge about the origins of tapeworms in humans, it is well known that certain tapeworms live in both animals and humans. Some tapeworms have such a complex life that they are required to live first in a herbivore (such as a cow) and then in a carnivore (such as a human) where only then can it reproduce.[30, 31, 32]

WHY DO MOST PEOPLE WANT TO VOMIT WHEN THEY SEE OR SMELL SOMEONE ELSE VOMIT?

(Asked by Andreas Pergantis of Thessaloniki, Greece)

The smell of vomit arises from the short-chain aliphatic acids such as butyric acid. Butyric is a fatty acid and is also found in rancid butter and parmesan cheese. In fact, butyric is from the Greek *butrous* which means 'butter'. Butyric acid is from animal and vegetable fat and remains liquid at room temperature. It has a rather disgusting aftertaste and a particularly nasty odour. So it is common for children to remark when they first smell parmesan cheese that it smells like vomit. Vomiting can be caused by many things ranging from food poisoning to gastritis, brain tumours to emotional upset. Drinking too much alcohol can cause vomiting. But regardless of the causes, it is in the best interests of everyone to avoid unnecessary contact with vomit, as it may very well be full of viruses, bacteria and toxins. The sight and smell of vomit trigger disgust and avoidance behaviour as a psychological response to the physical danger. Avoiding vomit improves our chance of survival. Many non-human primates share this behavioural trait with humans. In addition, vomiting can be social. If one person in a group vomits, it is advantageous in a survival sense for others in the group to be wary. The repulsion to sight or smell of vomit may help us survive. Evolutionary biology would suggest that this reaction is very deeply embedded in us.

WHY DOES OVER-EXERCISE MAKE US VOMIT?
(Asked by John Wright of London, UK)

There is at least one theory as to why vomiting occurs as a result of severe exercise. When the body is working hard, it may not have the capacity to also digest a large meal. It is too preoccupied with the strain of other body processes — such as cooling the body or sending blood to the extremities during marathon running — to utilise the meal. It is much easier to expel the meal. This is why running immediately after eating a large meal is so uncomfortable, and possibly dangerous. It is also why some people vomit after they eat a large meal while they are sitting in the hot sun. This may be part of the origin of the 'mother's advice' that you should not go swimming for an hour after you have eaten a large meal.

WHAT IS THE SIGNIFICANCE OF VOMITING DUE TO MORNING SICKNESS?

Vomiting during morning sickness reduces the likelihood of miscarriage, pre-term birth, low birth weight and death of the newborn. The theory holds that vomiting makes pregnant women avoid foods that have chemicals likely to damage the developing central nervous system of the embryo and foetus. From an evolutionary biological viewpoint loss of kilojoules due to food lost through vomiting is balanced by improved prevention of harm to the developing foetus.

Morning sickness usually ceases at the end of the first 3 months of pregnancy because by then the central nervous system of the foetus is less susceptible to the assault that a hazardous chemical could provoke. Statistically, between 6 and 18 weeks into pregnancy, women who experience morning sickness have fewer miscarriages than women who do not experience it. And women who vomit during morning sickness have even fewer miscarriages than those who are merely nauseous. So if you are a woman with morning sickness, it might make you feel a little better to know that it is good for your baby.[33]

◆

The science of vomiting is called emetology.

◆

The great English diarist, Samuel Pepys, reportedly buried his beloved parmesan cheese during the Great Fire of London in 1666.

◆

A study by Drs G. Koren and C. Maltepe[34] of 26 women with severe morning sickness found that anti-vomiting medications (antiemetics) taken 'before symptoms of morning sickness started appeared to prevent recurrence of severe morning sickness in subsequent pregnancies'.[35]

◆

A study by Dr C. Louik and colleagues[36] of 22,478 women found that the more pregnancies a woman experiences, the more likely she will experience morning sickness — a foetal 'dose' effect.[37]

HOW CAN HUMANS DIGEST TRIPE?
(Asked by Amy Grambs of Shreveport, Louisiana, USA)

Tripe is part of the diet of numerous human groups all over the world. Tripe is the stomach lining of animals such as cows. As the human stomach cannot digest its own lining, how can it digest the stomach lining of some other animal? The reason why the stomach lining does not digest itself is the reason why it can digest tripe. Our digestive system contains gastric acids and enzymes that break down organic materials, including the animal tissues of the meat we eat. Our own body tissues of the stomach lining would be digested by these gastric acids and enzymes were it not for a thick coating of mucus protecting our stomach lining. In tripe, the animal tissue is broken down by our stomach because the protection of mucus is no longer present in that animal's tissue, so it is just like any other meat we eat.[38]

◆

It takes food 7 seconds to get from your mouth to your stomach.

◆

The stomach has 35 million digestive glands.

WHAT DOES SALMONELLA HAVE TO DO WITH SALMON?
(Asked by Maurice Wills of Avalon Beach, New South Wales)

Salmonella has nothing to do with salmon. The name derives from Dr Daniel Salmon, the US microbiologist who identified salmonella in 1885. Salmonella is a general term applied to over 300 types of related bacteria found in food. Most cases, in fact, occur in the home and not in restaurants as some people think. But as a higher proportion of meals are eaten outside of the home in our modern, 24/7, post-industrial society, salmonella cases may increase. The effects of salmonella can vary, depending on the bacteria involved. Symptoms can range from a mild infection, such as stomach ache, to severe symptoms causing death.[39]

DO WE STILL REMOVE THE APPENDIX AS OFTEN AS WE USED TO?
(Asked by Peter Fletcher of Sydney, New South Wales)

As with circumcision and the removal of tonsils, we do not remove the appendix as often as we used to. Dr Dean Edell, the famous US radio and television physician, once said that the appendectomy paid off more swimming pool loans of doctors than any other surgical procedure. A generation ago, the slightest sign of an appendix problem and the doctor reached for the scalpel. No one seems to know for sure if this surgery is still undertaken as often now. It may differ from place to place. In one recent study, Dr I. Ahmed and colleagues[40] found that 'appendectomy ... is still practised by 75 per cent of general surgeons in the Mid-Trent region [of the UK] ...' Less than 25 per cent of surgeons employ some other form of treatment besides surgery in appendix cases. The doctors write that: 'At present, there is no agreed consensus on the management of such cases ... There is a need to develop a protocol for the management of this common problem.'[41, 42]

WHY IS THE BUTTOCK SHAPED THE WAY IT IS?
(Asked by Heather Norman of Geelong, Victoria)

Compared to humans, our primate cousins the chimpanzees and gorillas are 'buttockically' challenged. By contrast, our bottoms are larger and more rounded, curvy and better padded. In humans, the buttock muscles (gluteals) are stronger and far more massive than the puny ones of other primates. Our gluteals come in three sizes and types: small (*gluteus minimus*), medium (*gluteus medius*) and large (*gluteus maximus*). The size and strength of these muscles are both necessary for, and a consequence of, upright walking and running on two legs. Upright walking and running also demands stronger and larger calf and thigh muscles for pulling the legs backwards and forwards, and for getting and keeping the body moving. Chimpanzees and gorillas can only shuffle since they lack large and strong buttock muscles.

WHY DO WOMEN'S HIPS SWAY MORE WHEN THEY WALK?
(Asked by Heather Norman of Geelong, Victoria)

Women's hips sway more when they walk because they have a slight tilt in their pelvic girdle due to its larger size. The pelvic girdle is the bony ring formed by the hip bones and the sacrum — the bone to which the lower limbs are attached. Women have to have a larger pelvic girdle so that the head of an infant can fit through during childbirth. Quite a squeeze at the best of times! Wearing high heels emphasises the sway even more by tilting the pelvis a bit further forward.

WHY IS THE BUTTOCK LARGER IN WOMEN THAN IN MEN?
(Asked by Heather Norman of Geelong, Victoria)

There are many theories for this. One is that women need extra fat to give birth and breast feed babies. The buttock is a good place to store this extra fat. Another theory is from Timothy L. Taylor.[43] Taylor argues that an upright posture partly covered the vulva, presumably from behind. To compensate, women developed larger buttocks as a sexual signal. The larger the buttock, the larger the signal?[44]

181

WHY DO SOME AFRICAN WOMEN HAVE SUCH A LARGE BUTTOCK?
(Asked by Heather Norman of Geelong, Victoria)

The unusual storage of fat in the buttock is called steatopygia. When it is stored in the upper thighs it is called steatomeria. *Steat* is Greek for 'fat'. Prehistoric cave paintings indicate that women during the last Ice Age may have been more prone to both conditions. Today, these conditions are very common among the indigenous Andaman Islanders off the coast of India, as well as the Khoi people of South Africa.[45]

◆

The pelvic girdle is also called the *cingulum membri inferioris* or CMI.

◆

The *gluteus maximus* is the heaviest muscle in the entire human body.

Chapter 12
Otherwise Inside

WHAT LIES WITHIN: LIFE IN THE HUMAN BODY

The inside of the human body is alive with microscopic life of all kinds. There are at least 200 species of creatures that inhabit a healthy human body; even more when the body is infested with something. In his classic work, *Life on Man* (1969), Theodor Rosebury estimates that there are 80 distinguishable species living in the mouth alone and another 80 species living in the human gut. Among the bacteria living in the mouth are *Actinomyces viscosus* and *Actinomyces naeslundii* — together they constitute the sticky substance known as plaque that must be brushed away to avoid it hardening into tartar (calculus). The number of bacteria present varies with the part of the body. Rosebury estimates that the total number of bacteria excreted each day by an adult ranges from 100 billion to 100 trillion. Where there is a liquid flow, such as the tear ducts or the genitourinary tract, the number of bacteria is far less. In fact, Rosebury did not find any bacteria living in the bladder and in the lowest reaches of the lungs. Although the microbe density in such places as the human bowel is well into the billions of creatures, they take up little space. All the bacteria living in the human body would fill a container of only 300 ml.

If we are infected, inside the digestive tract is the protozoan called *Entamoeba histolytica* that causes amoebic dysentery and that can burrow through intestinal walls and into the bloodstream to infect the liver, lungs and brain. The digestive tract can also play host to hookworms such as *Ancylostoma duodenale* and *Necator americanus*, which measures about 10 mm and can also migrate via the bloodstream; a beef tapeworm called *Taenia saginata*, which can grow up to 20 m long; and the hermaphroditic *Shistosoma* worms, which can scar the bladder and make it bleed. The lymphatic system can sustain the *Wucheria* worm, which can be 12 cm long. The liver may let stay the bile-loving *Clonorchis sinensis* fluke; and the human brain can sustain the *Naegleria flowleri* amoeba. It loves the warmth of the brain and while squatting there continually reproduces in the millions until one day the person drops dead. Party's over![1, 2]

HOW FAST DOES IT TAKE THE BODY TO …?

It takes time for everything, and this includes everything that happens in the human body. Here are some examples:

- Fingerprints form 6 to 8 weeks before birth.
- Fingernails grow about four times faster than toenails — about 0.05 cm per week.
- At the time of birth, the human female possesses 400,000 egg cells in both ovaries. Of these, only about 480 may ovulate during her entire reproductive life. And of these, only 5 per cent or so will be fertilised. On average, it takes 72 seconds for a mature egg to be pushed out of the ovary. The fertilised egg remains within the oviduct for about 3 days before it enters the uterus.
- In the testes of the normal human male, a thousand sperm cells are produced every second. It takes about 2 months to manufacture a fully mature sperm cell. After ejaculation, the sperm swim for the egg cell at the speed of 15 cm per hour. This is the equivalent of a human swimmer covering about 12 m per second. Sperm reach the fertilisation site in about 50 minutes and remain alive for roughly 2 days.

- The amniotic fluid that surrounds the embryo and foetus during development is anything but a stagnant pool. While over 98 per cent of it is water, between 1 and 2 per cent is made up of substances such as foetal hair, skin cells, enzymes, urea, glucose, hormones and lipids. It is constantly and completely replaced about every 3 hours.
- In the foetal brain, nerve cells develop at an average rate of more than 250,000 per minute. At birth, a child's brain contains close to a trillion nerve cells. After birth, this rate of neuron growth slows down dramatically.
- Taste buds are among the earliest sense organs to appear in the foetus. By the third trimester of pregnancy, foetal taste buds are responsive to chemicals in the amniotic fluid. The life span of a taste bud cell is about 10.5 days.
- Twins are born, on average, 19 days earlier than single births. Their larger combined size stretches their mother's uterine muscles causing earlier contractions which push the twins out. Here's an old joke: A mother-to-be carrying twins doesn't give birth for 60 years! After ultrasound is invented, doctors check her out. They find two little old men inside saying over and over to each other: 'After you. No, after you.'
- A lactating mother produces about 1.7 litres of milk every day. Even so, if milk is not continuously removed from a mother's breasts, the ability to continue secreting milk is lost within 1 or 2 weeks. However, if the mother continues to have her breasts stimulated, milk production can continue for several years. Milk begins to flow within 30 seconds of an infant beginning to suckle. During this time, nerve signals move from the breast through the spinal cord and then to the brain. The brain then secretes the hormone oxytocin. Oxytocin travels through the bloodstream back to the breasts where it causes milk to be released.

- On average, nerve regeneration takes 4 to 6 weeks.
- Gums are renewed every 1 to 2 weeks.
- Eyelashes, which are more plentiful on the upper eyelid than on the lower eyelid, are shed continuously. Each of the more than 200 hairs per eye lasts from 3–5 months.
- The vibrations in the air constitute sound waves. The higher the pitch is the greater the frequency or cycles per second. Adults can detect sound waves that have a frequency between about 16 and 20,000 cycles per second. Yet they hear best at frequencies ranging from about 1000–2000 cycles per second. Children hear higher-pitched notes better than adults. After puberty, this sensitivity declines at the same time as the voice deepens. So our ears are best adapted to the pitch or sound frequencies of human conversation.
- The ability of the brain to detect the location of a sound depends on the differences in the time of the arrival of the sound to the two ears. We can detect the source of a sound even if it arrives in one ear 100th of a second before it gets to the other.
- The body loses water through the skin (from simple diffusion) at the rate of 500 ml per day. The body loses about the same amount of water each day from the lungs — in breath. A breath lasts about 5 seconds. Inhaling takes about 2 seconds, exhaling about 3 seconds.
- Nicotine, a component of tobacco smoke, gets into the body very rapidly. With one puff on a cigarette, nicotine reaches the brain in 7 seconds. This is several seconds faster than it takes alcohol to get into the brain. Even heroin, when it is injected into the arm below the elbow, takes twice as long as it does for nicotine to reach the brain.
- Statistically, one's life is shortened 14 minutes for every cigarette smoked.

- On average, a single cancer cell divides only once every 100 days. At that rate, owing to exponential growth (1 cell dividing into 2, 2 into 4, 4 into 8 and so on), 1 cancer cell may take 8 years to form a pea-sized tumour. But years later, the tumour will be about the size of a rockmelon and weigh more than 500 g.
- Human adult males have 4 or 5 erections per night while sleeping. Each lasts about 10 minutes, if uninterrupted.[3, 4]

DOES THE HUMAN BODY GENERATE LIGHT? DOES THIS ACCOUNT FOR THE BELIEF IN HALOS?

Ever wonder where the belief in halos comes from? Ever wonder why some Biblical accounts of Jesus Christ's resurrection describe light coming from his head, hands and feet? It is not widely known, but the human body generates light. Our hands, the bottoms of our feet, our foreheads, and especially our fingernails give off small but measurable amounts of light. According to Drs K. Nakamura and Mitsuo Hiramatsu,[5] the body shines light by releasing photons (tiny, energised pockets of light). They note that 'the presence of photons means that our hands are producing light all of the time'. Drs Nakamura and Hiramatsu have measured the photon output from the human body. In the hand, fingernails release 60 photons, fingers release 40 and palms release 20. The light is invisible to the naked eye although some individuals claim that they can see it. Some have even reasoned that the supposedly perceived light from the hands, feet and foreheads forms part of the basis of belief in halos. And if the resurrected Christ had light emanating from his forehead, hands and feet, these would be precisely the parts of the body in which the 'light spots' would come. Although Drs Nakamura and Hiramatsu do not know why fingernails have more photons, they have found that warmer temperatures increase the release of photons as does giving the subject more oxygen to breathe and rubbing the hands with mineral oil. The level of photon light seems to drop somewhat in unhealthy people. It is hoped that this line of research will one day assist medical diagnoses of diseases. It does give you something to think about.[6]

IS THERE LIGHT INSIDE THE HUMAN BODY?
(Asked by Jonathan Marten of Twickenham, Richmond upon Thames, UK)

A famous Groucho Marx punchline is 'inside of a dog, it's too dark to read'. While this is true, the human body is not entirely impervious to light. Light manages to permeate the skin and other tissues at least to some extent. But not surprisingly, the deeper into the body, the less light can be seen from outside. Interestingly too, the foetal environment within the womb is not entirely black — some light gets through the mother's abdomen. In fact, when a torch is shined on a mother's abdomen, a 6-month-old foetus will move towards the light!

IS IT POSSIBLE FOR YOUR ORGANS TO BE ON THE OPPOSITE SIDE OF YOUR BODY?
(Asked by Ian Hawthorne of East London, UK)

It is indeed possible to be born with some or even all of your organs on the wrong side of your body. The condition is called *situs inversus*. *Situs* is Latin for 'place' and *inversus* is from the Latin *invertere* meaning 'turn around'. *Situs inversus* literally means 'turned around in place'. The condition is surprisingly common. One in 8500 people have *situs inversus* involving at least one organ. For example, the heart is on the right side, the liver is on the left, etc. In 2007, Drs M.P. Guedj and A. Lev[7] reported on the case of *situs inversus* in a 62-year-old woman. She was scheduled for spinal surgery when it was discovered that her heart was reversed. Drs Guedj and Lev write that the diagnosis of *situs inversus* was 'obvious'. They recommend that in such cases, doctors should 'leave well alone'.[8]

WILL THE HUMAN BODY EXPLODE OR FREEZE, OR WILL BLOOD BOIL IF AN ASTRONAUT IS EXPOSED TO THE VACUUM AND ZERO GRAVITY OF OUTER SPACE?

(Asked by Roger Sutton of Camarillo, California, USA)

In the film *Total Recall*, Arnold Schwarzenegger leaves his protective enclave on Mars and is exposed to the hostile Mars environment. He swells up like a balloon but is saved. Similarly in the film *Outland*, Sean Connery is an outer space construction worker who has torn his spacesuit and it begins to leak. As the internal pressure drops he is exposed to the vacuum, he swells up and finally explodes. In the classic film, *2001: A Space Odyssey*, Keir Dullea, without a helmet, 'blows' himself into the airlock from his space pod without blowing up. Despite these horrific and dramatic images in popular films, if there was a problem with an astronaut's spacesuit, the most trouble they would experience would likely be difficulty breathing. Air pressure would drop and the human diaphragm could not pull hard enough for normal respiration. It would become harder and harder to inhale, an anoxia (no oxygen) condition would result, then unconsciousness and then death. Much of what the unfortunate astronaut would experience in outer space would be similar to what an unfortunate deep sea diver would experience on Earth. Dissolved nitrogen in the blood would bubble out causing caisson sickness — also known as decompression sickness or 'the bends'.

Caisson sickness is caused by rapid decompression. It can appear in workers building tunnels or supports for bridges within enclosed units under high atmospheric pressure designed to keep surrounding water out, called caissons. Ironically, you can experience the same effect high up in outer space or low down under ground. According to NASA: 'If you don't try to hold your breath, exposure to space for half a minute or so is unlikely to produce permanent injury ... You do not instantly lose consciousness ...' Various problems such as eardrum damage, sunburn and tissue swelling may occur after about 10 seconds of exposure. Normal blood pressure keeps the blood from boiling. If anything, saliva may be more likely to boil. NASA adds: 'You do not explode and your blood does not boil because of the containing effect of your skin and circulatory system. You do not instantly freeze because, though the space environment is typically very cold, heat does not transfer away from a body quickly. Loss of consciousness occurs only after the body has depleted the supply of oxygen in the blood. If your skin is exposed to direct sunlight without any protection from its intense ultraviolet radiation, you can get a very bad sunburn.'[9] There have been some incidences in which astronauts have been briefly exposed. In one training exercise in a vacuum chamber, un-oxygenated blood reached the astronaut's brain and he lost consciousness after 14 seconds. According to Dr Samuel Conway,[10] lack of oxygen is the big problem way out there! Evaporation in a vacuum occurs quickly. Any cuts in the skin would quickly seal with dried blood. You wouldn't bleed for long as you wouldn't breathe for long![11, 12]

WHAT HAPPENS TO FEMALE ASTRONAUTS WHEN THEY MENSTRUATE IN SPACE?
(Asked by Richard Pierce of Lincolnshire, UK)

Menstruation is not an issue in space travel. Gravity is not essential for menstruation to occur. Menstruation is a very complicated physiological process involving the internal factors of many different hormones, the woman's sexual organs and the brain. Very little blood is lost during menstruation. So it is not considered a major 'waste management' problem by space flight scientists. When menstruation occurs in a zero gravity environment it would have to be dealt with somewhat differently hygienically. But to do so is far less of a challenge for scientists than the far more important 'waste management' problems posed by urination, defecation, infections and a few other normal body processes and events that can be expected when humans travel in space. Drs Richard Jennings and Ellen Baker[13] point out that 'there currently are no operational gynaecological or reproductive constraints for women that would preclude their successful participation in the exploration of our nearby solar system'.[14] Dr Baker is also an astronaut with the NASA–Johnson Space Centre in Houston.

Women have been an important part of space crews since almost the beginning of space travel in 1961. There are gender differences in the ability to withstand extremes, such as reduction in oxygen supply (hypoxia), heat, cold, decompression, acceleration, isolation and 'impact'. However, as Drs Jennings and Baker observe, these differences are generally minor, often depend on acclimatisation and individual variation, and favour women as often as men.

◆

The first woman in space was Valentina Tereshkova who flew on the Soviet Union's Vostok 6 flight in 1963. In the US space program, the first woman in space was Dr Sally Ride in 1983.

◆

The vast majority of women astronauts have not had children. Most delay their first pregnancy until their space career is over. Of these women, the average age of giving birth for the first time is 40.

◆

It is certainly possible for humans to conceive in space. Such experiments were rumoured to have taken place in some Soviet space flights, but were officially denied at the time.

◆

NASA will not allow women astronauts to participate in neutral buoyancy training while pregnant. This involves dives underwater lasting up to 8 hours. The pressure changes are thought to be harmful to the developing baby.[15, 16]

CAN YOU CATCH A COLD IF YOU GO OUTSIDE ON A COLD DAY WITH WET HAIR?
(Asked by Miyoki Takahashi of Tokyo, Japan)

There is no clinical evidence that if a healthy person has a short exposure to a draught or goes outside with wet hair, this will cause them to catch a cold. Colds are spread by more than 200 different viruses. Temperature change up or down does not by itself cause a cold. While asthmatics and others with chronic respiratory problems may experience shortness of breath or coughing in a cold environment, even they are not more susceptible to catching a cold. Neither are people with lupus, HIV or rheumatoid arthritis more at risk.

According to physician and media doctor Dr Gabe Mirkin: 'the real question about colds is whether chilling the body hinders your immunity so that you can't kill the germs in your body ... so the germs that you can normally control suddenly become pathogens and make you sick, because your immunity is suppressed by you being cold.' Dr Mirkin adds: 'Chilling does not hinder your immunity as long as you are not so cold that your body defences are destroyed.' One of the first of many studies demonstrating this was undertaken in the 1960s in a Texas prison. Inmates had a virulent cold virus placed directly into their noses. At varying times after their exposure to the virus, they were exposed to extreme temperatures, with varying amounts of clothing. It was found that being cold or warm, dressed or undressed, or having wet hair or dry hair had no effect on their infections.[17] (Note: Ethical considerations of today would probably not allow such a study.)

CAN YOU CATCH DIARRHOEA, BECOME CONSTIPATED OR DEVELOP HAEMORRHOIDS BY SITTING FOR TOO LONG ON A COLD SURFACE?
(Asked by Miyoki Takahashi of Tokyo, Japan)

This seems to be a popular myth. A change in bowel movement, such as diarrhoea or constipation, or the developing of haemorrhoids is not produced by sitting for too long on a cold surface. No matter how cold and no matter how long. According to Dr Terry Bolin:[18] 'Sitting on cold ground has nothing to do with it.'[19]

DO YOU REALLY HAVE TO WAIT AN HOUR AFTER EATING BEFORE YOU GO SWIMMING?
(Asked by Bianca Rossi of Bendigo, Victoria)

There is only a small degree of truth in the advice to wait an hour after eating before going swimming. According to Dr Andrew Lloyd:[20] 'This advice is only applicable if you are going scuba diving after eating. If you are holding your breath, this could cause pressure on your stomach and if your stomach is full this could cause reflux.'[21] According to Dr Richard Fedorak:[22] 'The simple average meal isn't going to affect your ability to get into the water. That's a myth, and we need to myth bust.'[23] Statistics show that less than 1 per cent of all drowning deaths in the US occurred after the victim ate a meal or where eating a meal was ruled a factor in the case. The American Red Cross makes no recommendation about waiting any amount of time after eating a meal before swimming.

It was once believed that oxygen in the blood diverted to the stomach to help with food digestion was pulled away from the extremities, thus resulting in muscle cramps. However, the human body is such that plenty of oxygen in the blood is left over to 'feed' the extremities even when a large meal is being digested. According to Dr Roshini Rajapaksa:[24] 'Swimming strenuously on a full stomach could conceivably lead to cramps, but for most recreational swimmers the chances are small.'[25] Nevertheless, it seems common sense would dictate that if you have just eaten a large meal, do you really want to exert yourself by *any* kind of heavy exercise?[26]

WHAT IS THE PROSTATE, AND WHAT DOES IT DO? WHY DON'T WOMEN HAVE ONE, AND DO WOMEN HAVE AN EQUIVALENT?

(Asked by Luigi Imperiale of Rome, Italy)

The prostate is an exocrine gland in the human male that surrounds the neck of the bladder and urethra. A healthy prostate is about the size and shape of a walnut. 'Exocrine' means it secretes into a duct. The prostate contributes to and stores a clear fluid that makes up as much as 30 per cent of the volume of male semen. It also has some muscles that help advance the semen out through the urethra. The prostate is located just in front of the rectum, which is why the prostate can be felt in a rectal exam. The bladder is on top of the prostate, so when it fills it must empty. The urethra is the tube connecting the bladder to the outside of the body. Think of it this way: if the bladder is the cistern, the prostate is the faucet, and the urethra is the hose.

During embryonic development, the prostate actually forms from a portion of the urethra. The urethra is present in both males and females. The amount of male or female hormones determines whether or not a prostate is developed. In males, testosterone causes numerous tissue pockets (outpouchings) or tissue foldings (invaginations) to project out from the urethra into the surrounding tissue. This ultimately results in the prostate. In females, without as much testosterone, a much smaller pocket-forming and tissue-folding process occurs. This forms in females the wonderfully titled 'paraurethral gland of Skene'. There are two of these tiny glands in the human female, which are located on either side of the urethral opening. But they are so tiny that they usually cannot be seen without a magnifying glass unless they become infected and enlarged, cause pain or even leak a little fluid. If the prostate is the size of a walnut, the paraurethral gland of Skene is not even the size of a peanut.

◆

A common mistake is confusing the word 'prostate' with 'prostrate'. Prostrate means 1) stretched out, lying flat, with your face on the ground; 2) being overcome, lacking vitality, will or the power to rise; or 3) trailing on the ground. Ironically, all of these could happen to a man if he has prostate problems.

WHAT IS THE USE OF THE HYMEN? IF IT HAS A USE, WHY DOESN'T IT GROW BACK?

(Asked by Marc Savage of Northampton, UK)

The hymen (or maidenhead) is a ring of tissue around the vaginal orifice. Hymen is Greek meaning 'virginal membrane' or 'thin skin'. Hymen was also the Greek god of marriage. The hymen is shrouded in myth, one of which is that the hymen completely occludes the vaginal orifice in human females until the first sexual penetration. But this is quite rare. Another myth is that the human hymen is the same in all females, however there is much variation. There are in fact four distinct forms of hymen — annular, septate, cribriform and parous introitus — based on its general configuration. The many variations in the 'normal' hymen are very important for doctors to understand. Otherwise, they risk, among other things, the misdiagnosis in a child sexual abuse case. This is a point made recently by Dr A.K. Myhre and colleagues,[27] who point out that variations in the hymen are so great that a small percentage of female babies are born without one.[28] Yet another myth is that a so-called 'intact' hymen ensures that a woman is a virgin. But the hymen may be 'broken' for a variety of reasons besides sexual intercourse. Another myth is that the hymen is necessarily 'intact' at some time. In fact, there is always an opening in it of some kind. Still another myth is the blood that flows after the first sexual penetration is caused by the tearing of the hymen. In fact, blood is not always produced and if it is it may be from the tearing of surrounding tissue, not necessarily the hymen.

In recent years, the very concept of a hymen has been criticised, and even its very existence has been questioned by researchers. Many gynaecologists and other experts on the human female reproductive system consider most commonly held beliefs about the human hymen to be based more on cultural perceptions and sexual stereotypes than on anatomical and physiological reality. The hymen has no definite function other than as part of the vaginal opening. It may have a slight preventative function with respect to infection in infants. But this is disputed. The hymen does not 'grow back', although it can be surgically 'restored' according to some surgeons. What is not a myth is that the hymen has, and continues to have, great symbolic significance as an indicator of a woman's virginity in many religions and cultures throughout the world.[29]

WHAT IS MERLIN'S DISEASE AND CAN IT SHRINK MY BREASTS?
(Asked by Hailey Goode of Alexandria, Virginia, USA)

We are now in the realm of urban myth, because there is no such thing as 'Merlin's Disease' — certainly nothing that is going to shrink anyone's breasts. Some years ago a tabloid newspaper ran a story about Merot's Syndrome[30] which, supposedly, could melt down large breasts to the tiny size of mosquito bites in no time at all. Huge headlines warned of this new, so-called 'Boob-Onic Plague'. Merot's Syndrome was allegedly caused by bacteria that feed on the body fat generated by the female hormone oestrogen. It was reassuringly pointed out to readers that since Merot's Syndrome attacked only natural breast tissue, 'Pamela Anderson's prize assets are safe'. Merot's Syndrome was claimed to be highly contagious. It was transmitted by breast-to-breast contact (even through clothing). Among other things, women were warned against borrowing a halter top or bra from a friend, sharing a sauna or hot tub with a flat-chested woman, or squeezing a stranger's breasts to see if they're real 'even if invited to do so'. Instead, the advice was 'just take her word for it'. Lest the same fate befall the unwary, women were also warned to avoid contact with the 'bosomly-challenged' in a crowded bus, subway or elevator. In particular, cheerleaders were warned to be vigilant.

Merot's Syndrome was allegedly named after Dr Jean-Pierre Merot, 'the acclaimed French physician who first isolated the antibiotic-resistant bug'. Now the problem with all of this and what puts this into the domain of urban myth is that the scientific literature has no record of 'Merot's Syndrome', no record of an 'acclaimed French physician' by that name, and no record of any disease that shrinks breasts in such a strange manner. Breasts may reduce in size over time and for many well-established reasons, but not by Merot's Syndrome — or 'Merlin's Disease'. The legendary Merlin was the wizard of King Arthur's legendary court at Camelot — all mythic. The same may be said for 'Merlin's Disease'.

WHY DOES STRETCHING FEEL SO GOOD?
(Asked by Tony Ferenczi of New York, USA)

When muscles are not worked very much, the flow of blood through them is reduced. Oxygen levels drop and metabolic wastes such as carbon dioxide accumulate. When you stretch, this oxygen-poor blood is squeezed out of the muscles and replaced with fresh oxygen-rich blood. The body craves oxygen and getting more to the muscles feeds this craving — and feels so good. A more complicated answer is that muscles are made up of protein filaments which are composed of actin and myosin molecules. These molecules are continually interacting and linking at varying speeds depending on, among other things, the level of muscle activity. In an active muscle, the interacting and linking of actin and myosin molecules is hearty and vigorous; in a relaxed muscle, the actin and myosin activity is slowed down, but never entirely stopped. This muscle inactivity generates a molecular activity that creates small but significant resting tension and muscular stiffness. When you stretch a muscle, you activate it — like waking it up. A force is generated by the combined action of many interacting actin and myosin molecular linkages. In particular, large stretches temporarily break the molecular cause for the muscle tension and the stiffness. A stretch gives almost instant pleasure although the resting tension and stiffness do not arise instantaneously. Your body loves all the stimulating molecular activity going on — and it feels good.

This is an example of the principle of 'stationary rigidity' (thixotrophy) at work. Named by Dr Derrick Denny-Brown in 1929, it was found that there is muscle 'reluctance' to move, hence 'stationary rigidity'.[31] The degree of stationary rigidity is proportional to the length of time the muscle has not moved. The longer the muscle has not moved, the greater the 'reluctance' to move. The more the muscle moves, the greater the 'willingness' it has to move some more. This principle is behind why athletes train to stay in shape, why a massage feels both relaxing and invigorating, and also why stretching relieves stiffness and feels so good. The body is not the only thing that is thixotropic — so is tomato sauce in a bottle.[32]

◆

You use 300 muscles to balance yourself when standing still.

◆

Fidgeting can burn as many as 1470 kilojoules per day.

◆

'Use it or lose it!' is true when it comes to muscles. But it takes twice as long to lose new muscle tissue if you stop working out as it did to gain it.

IF THE BODY CONTAINS IRON, IS IT POSSIBLE FOR PEOPLE TO BECOME MAGNETISED?

(Asked by Nikki Hunt of Burnaby, British Columbia, Canada)

There's so little iron in the human body that there is no chance of it becoming magnetised. According to Dr Michael Onken[33] the American Red Cross specifies that 500 ml of blood contains about 250 mg of iron.[34] The average human has about 4.7 to 5.6 litres of blood in their body which equates to about 5–6 g of iron. This is about the weight of a US 5-cent piece. Measured against the body mass of the average person, the amount of iron is far too small to get anywhere near being able to cause the body to be magnetised. Although the body contains only a tiny amount of iron, that small quantity is essential for carrying oxygen in the blood and using oxygen in cells.[35]

◆

Your body contains about the same amount of iron as in an average iron nail.

WHY DO SOME PEOPLE FEEL THE COLD MORE THAN OTHERS?

(Asked by Mick Higgins of Toronto, Ontario, Canada)

Human body temperature is a measure of the body's ability to keep, generate or get rid of heat as the need arises. The body is very adept at maintaining temperature within a narrow and safe range despite occasional huge variations in the surrounding temperature. But some bodies are more efficient than others. Even bodies of the same height and weight may differ dramatically in the ability to maintain body temperature. Humans also differ in their preferred room temperature. Some like it warmer, some cooler. This is called thermal comfort, which is not merely physical but psychological too. In one's choice of preferred temperature, besides psychology, other personal factors come into play. Two of these are clothing levels (their insulation value) and activity level (their effect on body metabolism). According to Stephen Turner:[36] 'Activity level is a key factor, as illustrated by the story of an adult skiing down a mountain with a child bundled in warm clothes on their back. At the bottom of the hill, the adult is sweating, but the child is bitterly cold. Difference in metabolic rate is the cause.'[37]

The American Society of Heating, Refrigerating and Air-Conditioning Engineers in Atlanta, Georgia, maintains and periodically revises international thermal comfort standards. But this is not the end of the story. According to Paul Spry, a consultant engineer in Canberra, there is also evidence that a person's thermal comfort perceptions are influenced by external factors such as recent weather. He adds: 'Another consideration is the difference between sensory and psychological reactions. A particular person may say it is cold or cool, but if you accept this as the final judgment (as a desire for a higher temperature) you are at risk. Further questioning may reveal that though the person is "cold" or "cool" they may be optimally comfortable (desire no change for thermal comfort reasons).'[38]

◆

When you are hot, the blood vessels in your skin expand (dilate) to carry the excess heat to your skin's surface where it is released. You sweat and as the sweat evaporates, it cools the body. When you are cold, the blood vessels in your skin narrow (contract) so that blood flow to your skin is reduced to conserve body heat. You may start shivering, which is involuntary, and experience rapid contraction of the muscles. This extra muscle activity helps generate more heat. Both processes help keep your body temperature within the required narrow and safe range.[39]

◆

An article in *Ergonomics* (August 1994) concludes, among other things, that wet underwear has a 'demonstrated' cooling effect on the body.[40, 41]

◆

The body gives off enough heat in 30 minutes to bring 1.5 litres of water to a boil.

❖

In a hot climate, you can sweat as much as 500 ml of water per hour.

WHAT IS A FISTULA?

(Asked by Terry Cardner of St Louis, Missouri, USA)

The body grows all sorts of things. Parts of the body are connected in all sorts of ways. A fistula is an abnormal connection, such as a tube, between an organ, a vessel or an intestine and another organ, vessel or intestine. The connection can also involve the skin. Fistulas are usually the result of injury or surgery, but can also come from infection or inflammation. Anyone can develop one. Inflammatory bowel disease, such as ulcerative colitis or Crohn's disease, is an example of a disease that promotes the development of fistulas. In such a case the fistula is between one loop of intestine and another. Fistula is Latin for 'pipe' or 'long narrow ulcer'. Dr J.M. Draus and colleagues[42] write that fistulas can be very serious. In their study of 106 patients with enterocutaneous fistulas 7 per cent died of complications. But for most such fistula patients, there are very satisfactory treatments. The Draus team points out that 75 to 85 per cent of patients are healed after surgery. See your doctor if you have any worries in this area.[43, 44]

DOES DRINKING ALCOHOL REALLY KEEP YOU WARM?

Although long disproven, drinking alcohol to keep warm is one of those myths which still hangs around. A strong 'No!' comes from the Swiss Institute for the Prevention of Alcohol and Drug Problems in Lausanne. If the legendary St Bernard finds you stranded in the icy Alps, you would be better off hugging the hound than downing the hooch of the pooch. Alcohol only gives a false sense of warmth, but the dog could pass along some lifesaving body heat. Even a little nip from the brandy keg will send your blood to the surface of your skin. You may feel warmer, but your blood will actually be cooled. Many heat-sensing nerves are located near the surface of the skin. Drinking can make you temporarily feel warmer. In fact, while you get the feeling of warmth from alcohol, it is really unsuitable because it allows the cold to enter the body.[45]

DOES DRINKING ALCOHOL THIN YOUR BLOOD?

Alcohol does not thin or thicken blood. Alcohol is a vasodilator and causes the blood vessels to expand. This is particularly true for the tiny capillaries located just below the skin's surface. The normal thermostatic control of the body is altered by alcohol ingestion. The blood vessel dilation allows a greater amount of blood volume to be brought to the skin's surface. This facilitates heat loss and also explains why your face looks flushed when you have been drinking. But alcohol does not thin blood.

WHY DOES DRINKING ALCOHOL MAKE YOU FEEL THIRSTY?

Alcohol ingestion forces the body to metabolise it in order to remain chemically balanced for proper body functioning. In doing so, the body actually draws water from body tissues. This can cause a thirsty feeling, and drinking more alcohol only makes it worse.

IS ALCOHOL A BIG FACTOR IN ACCIDENTS?

Alcohol is a major factor in accidents with injuries. According to Drs Gerhard Gmel and Jurgen Rehm:[46] 'The research evidence indicates a high level of alcohol involvement in all types of unintentional injuries. The number of drinks consumed per occasion, especially when indicated by BAC [blood alcohol concentration], is strongly associated with the occurrence of injuries, independent of the usual frequency and quantity of alcohol consumed. Drinking may be less associated with workplace injuries for various reasons, but appears to play a role in causing falls, the second most common form of unintentional injury.'[47]

◆

After drinking alcohol, the last place in the body to be cleared of the effects of it is the brain. This is why we are often clumsy, stumble and are slow to react when drunk. Our brain has not caught up with our body.

◆

Even mild dehydration will slow down the body's metabolism by as much as 3 per cent.

◆

The thirst mechanism is so weak that it is often mistaken for hunger.

◆

Between 30 and 75 per cent of people are chronically dehydrated, especially among the elderly. As the population ages, the rate of chronic dehydration goes up.

Chapter 13
Behaviour

WHAT IS THE DIFFERENCE BETWEEN PSYCHOLOGY, PARAPSYCHOLOGY AND PSYCHIATRY?
(Asked by Angela Cross of New York, USA)

Psychology is the name given to the academic discipline and science of human behaviour through the investigation of the mind and experience. It is the study of people and explores their individual behaviours, including thoughts, feelings, actions and experiences. Psychologists want to know what guides and drives a person. They want to understand all about how people act, react and interact. *Psykhe* is Greek for 'the soul'. It was first used by Rudlof Goeckel (1547–1628) in connection with the study of the soul, not the mind. The usage expanded to mean 'brain' with the work of anatomist Thomas Willis in 1672, and to further include the 'mind' with the work of philosopher Christian Wolff in 1732 and 1734. It greatly expanded in the 19th century, distancing itself from philosophy, and even more so in the 20th century.

Parapsychology is the name given to the academic discipline and science of psychical research. *Para* is a prefix in Greek meaning 'alongside, beyond, altered and contrary'. As such, parapsychology exists alongside, beyond, altered and contrary to psychology. This suggests why parapsychology, compared with psychology, has had a greater difficulty being recognised as an academic discipline and as a science. Parapsychology can best be thought of as an adjunct to psychology. It explores questions that psychology often chooses not to ask. Concepts such as extrasensory perception or ESP (the ability to acquire information by means other than the five senses) and psychokinesis or PK, sometimes called telekinesis or TK (the ability of the mind to affect matter outside of the body).

Psychiatry is the name given to the branch of medicine dealing with the diagnosis and treatment of psychological and physical problems associated with mental or brain illness. *Psykhe* along with the Greek *iatreia* which means 'healing care' gives us our modern meaning of psychiatry as 'healing care for the psyche'. Psychiatrists employ talk, drugs and surgery in the performance of their therapy.[1]

WHAT IS THE DIFFERENCE BETWEEN A PSYCHOPATH AND A PSYCHOTIC?
(Asked by Kelli Pritchard of Dearborn, Michigan, USA)

Although 'psychopath' and 'psychotic' are words often used interchangeably, there is a distinction.

'Psychopath' is the designation for any individual who suffers from psychopathology — an anti-social type of personality disorder. The word itself is from the Greek *psykhe* meaning 'mind' and the Greek *pathos* meaning 'disease'. 'Psychopath' literally means one who suffers from a 'disease of the mind'. The English doctor James Cowles Prichard (1786–1848) is credited with introducing the term 'psychopath' into the fledgling field of psychiatry in his book, *Treatise on Insanity*, in 1835. But he meant the concept as a moral disorder and not in the anti-social personality sense that we use it today. The characteristics of a psychopathic personality are:

- Superficial charm and average or above average intelligence
- Absence of delusions and other signs of irrational thinking
- Absence of nervousness or psychoneurotic manifestations
- Unreliability
- Untruthfulness and insincerity
- Lack of remorse or shame

- Inadequately motivated anti-social behaviour
- Poor judgment and failure to learn from experience
- Pathologic egocentricity and an incapacity for love
- General poverty in major affective reactions
- Specific loss of insight
- Unresponsiveness in general personal relations
- Extreme and uninvited behaviour with or without drugs or alcohol
- Suicide threatened but rarely carried out
- Sex life impersonal, trivial and poorly integrated
- Failure to follow any life plan
- Manipulative approach to interpersonal relationships
- Glibness, superficiality, callousness, irresponsibility, irritability, impulsivity, low frustration tolerance, proneness to aggressive behaviour, arrogance and deceitfulness.

'Psychotic' is the designation for any individual who suffers from a psychotic disorder or psychosis. The word itself combines the Greek *psykhe* and the Latin *osis* meaning 'abnormal condition'. Psychotic literally means one who suffers from an 'abnormal condition of the mind'. Psychotic was used first in 1847 to refer to 'mental derangement'; in 1890 to refer to 'neurotic/neurosis'; and in 1910 to mean 'a psychotic person' in the sense that we use it today.

A psychotic disorder is one in which there is a gross impairment in reality testing. The individual incorrectly evaluates their own perceptions and thoughts, and makes incorrect inferences about reality even in the face of contrary evidence. They suffer from delusions, hallucinations, and sometimes both. Delusions are false beliefs regarding the self or persons or objects outside of the self that persist despite the facts. Hallucinations are subjectively experienced sensations in the absence of actual appropriate stimulus but which are regarded by the individual as real. The terms psychopath and psychotic are close in meaning so it is easy to see how they could be confused.[2, 3]

WHY DO SO MANY OF US BELIEVE IN CONSPIRACY THEORIES?

(Asked by Mike Pell of TV program *Weekend Sunrise*, Seven Network, Sydney)

In our global village, we respond to shocking news events in many ways including trying to make sense of them in emotional, moral, spiritual, political and scientific terms. When such an event happens such as President John F. Kennedy's assassination, Princess Diana's death or 9/11, we look for its explanation and meaning. When the explanation is not clear, the meaning is not clear, so confusion arises to compound the emotional upset. Conspiracy theories are satisfying since they place events in an understandable context and help us deal with events intellectually and emotionally.

Conspiracy theories emerge when flawed logic combines with evidence that is lacking, disputed or contradictory. This can be exploited by mythmakers, mischief makers and attention-seekers. But genuine researchers, investigators and truth-seekers may offer explanations for the event. The situation is not helped when governments withhold or fabricate information for 'reasons of state'. All governments keep secrets. Often a government's explanation for an event in itself can become a conspiracy theory, such as 9/11.

Once a conspiracy theory is formulated, it can continue through the momentum of a 'confirmation bias'. This occurs when the believer accepts all evidence that confirms the theory while rejecting all that does not. According to Dr John Gartner:[4] 'Conspiracy thinking is embraced by a surprisingly large proportion of the population. Sixty-nine per cent of Americans believe President John F. Kennedy was killed by a conspiracy ...'. Sixty per cent of UK adults believe that Princess Diana was murdered. Sixty per cent of US adults believe that the US government is withholding information about 9/11. According to Gartner, '36 per cent of respondents to a 2006 Scripps News/Ohio University poll at least suspected that the US government played a role in 9/11'.[5]

ARE CONSPIRACY THEORIES EVER GOOD FOR YOU?

According to some in the field of evolutionary psychology, conspiracy theories can be good for you. It is argued that paranoid tendencies are associated with an animal's ability to recognise danger. Higher animals attempt to construct mental models of the thought processes of both rivals and predators in order to read their hidden intentions and predict future behaviour. This ability is valuable in sensing and avoiding danger in the animal community. But so far there has been no gene found for such 'alertness'.

ARE THERE TYPES OF CONSPIRACY THEORIES THAT ARE MORE LIKELY TO BE BELIEVED?

In general, with all of the ingredients necessary for a conspiracy theory present, the more tragic the outcome of an event the more likely people are to believe it was the result of a conspiracy. This is the view of Dr Patrick Leman[6] in a paper presented to the British Psychological Society in 2003.[7]

Dr Leman also found that 'if people become distanced from institutions of power and state, they are more likely to distrust official accounts'.[8]

WHY DO WE DOODLE?
(Asked by Leanne Ward of New York, USA)

Doodles! Often we do it. Often we don't realise it. Often when caught doing it, we are embarrassed to have to explain it. Surprisingly these nonsense scribbles we leave behind on notepads, paper margins, desktops, walls and anywhere else where a pen can leave a trace may have meaning. In fact, for one US psychologist, doodling and what doodles mean has been the chief focus of his lifetime work. Dr Robert

Burns[9] studied doodles and used them to diagnose emotional problems of clinical patients. Dr Burns and other clinicians and researchers in the field of behavioural art therapy, maintain that the shapes and symbols we draw can reveal much about our state of mind. As Dr Burns is quoted as saying as far back as 1991 'even the most innocent doodle may carry messages from the unconscious'.[10] For example, a commonly drawn doodle is a tree. Trees represent growth and life. A full, leafy tree with a wide trunk suggests someone who is vital, energetic and with a strong will to live. Very narrow trees with leafless branches often appear in the drawings of the frail elderly, indicating that their spirit, their will to live, may be waning. According to Dr Burns: 'If you find yourself doodling pictures of houses, you probably place a high value on shelter and security.' Other symbols too are strong indications of things an individual values most, for example, numbers and dollar signs indicate a preoccupation with money; planes, cars, ships and other vehicles may indicate a desire to travel, alter relationships, or change one's life, and the list goes on. Ladders can be symbols of tension and precarious balance. Light bulbs and images of the sun suggest feelings of warmth and light. Squares, triangles and circles are the sign of a logical, analytical mind. None other than Sigmund Freud (1856–1939) and Carl Jung (1875–1961) are two pioneers of symbol interpretation in psychology — the co-fathers of 'doodleology' if you will.

According to Dr Burns, the analysis of doodling should be part of the clinical procedures of every psychologist or psychiatrist. He notes that 'messages are there, after all. No one's surprised that an electroencephalogram can chart brain waves using a stylus attached by way of electrodes to the brain. The only difference with doodling is that we use a pen attached to the brain by nerves and muscles.' Doodling becomes a kind of visual free association, a way of tapping the deep reservoir of self-knowledge contained not in words but in images. But what of the claim that the study of doodles is unimportant since there is no way of knowing what the scribbles symbolise to the person doing the scribbling — if anything at all? Dr Burns counters that it is only after careful study of doodles over many years and from many different individuals that the patterns of doodle symbolism and their significance emerge. He adds that 'even at their simplest, the idle jottings we repeat in the margins of our notebooks can evoke childhood memories and associations that provide clues even to our obsessions. Stars, for instance, show up all the time in the drawings of emotionally deprived children. Stars are what we wish upon. People who fill their doodles with stars may be longing for something they were deprived of, like love or affection.'[10] So take it from this doodle dandy — your doodles do mean something.

DO MEN AND WOMEN DOODLE DIFFERENTLY?
(Asked by Leanne Ward of New York, USA)

Dr Burns claims that gender is a factor in doodling patterns. He maintains that 'men tend to doodle geometric shapes while women are more likely to doodle human figures and faces. Physical features, especially any that are abnormally large or small, carry special meaning. Very large eyes suggest vigilance, for instance, or in extreme cases, paranoia. Very small eyes or no eyes at all, suggest someone who doesn't want to see. Long arms symbolise reaching out. An absence of arms means withdrawal.'[10]

DO OUR DOODLES REVEAL SEXUAL THOUGHTS?
(Asked by Leanne Ward of New York, USA)

Dr Burns asserts that one's relative preoccupation with sex also shows itself through one's doodles. Dr Burns observes that 'a preoccupation with sexuality usually shows up in figures whose genital areas are emphasised and heavily shaded or in the repeated use of classic sexual symbols such as snakes, candles, or darts striking a target.'[10]

◆

In January 2005, doodles found on the desk of British Prime Minister Tony Blair at Number 10 Downing Street in London were discussed by psychologists and handwriting experts as to their meaning. According to the BBC, newspaper stories contained phrases such as 'struggling to concentrate' and 'not a natural leader'. It then emerged that a mistake had been made. The doodles were in fact drawn by a visitor — US computer magnate Bill Gates.[11]

◆

The doodles and notes of US president John F. Kennedy are released periodically by the John F. Kennedy Library and Museum in Boston. A total of 135 pages were released in 2004. Subsequent historical events, long after JFK's assassination in 1963, give one JFK doodle a special significance: on one page a small circle was written with the numbers '9/11' contained within. Just to the lower left on the same page is the word 'conspiracy' — and it is underlined. Conspiracy buffs, take note![12, 13, 14]

◆

Famous doodlers include Erasmus (comical faces), John Keats (mostly flowers), Abraham Lincoln (various shapes) and Ralph Waldo Emerson (various shapes).

◆

Doodling boosts memory.[15] Experiments by Dr Jackie Andrade[16] indicate that during boring events those who doodle recall information better than those who do not. Dr Andrade suggests that doodling be required at business meetings — most of which are very boring.[17, 18]

WHAT IS A 'CONFESSING SAM'?

(Asked by Casey Filocamo of Edmonton, Alberta, Canada)

'Confessing Sam' is the term in criminal psychology for a person who makes a false confession after a particularly widely publicised crime has taken place. Some Confessing Sams will confess to just one infamous crime reported in the media; others will confess to *every* infamous crime. Confessing Sams will often continue to maintain their guilt long after police rule them out as suspects. The first genuine Confessing Sam was Robert Hubert. In 1666, the Great Fire of London destroyed 80 per cent of the city. Hubert confessed to having started the fire by throwing a crude fire grenade through an open bakery window. At his trial it was proven that Hubert, a sailor, had not arrived in England until 2 days after the fire started, therefore he could never have been near the bakery where the fire broke out. On top of which he was so badly crippled that throwing anything was beyond him. If that were not enough, the bakery had no windows. Nevertheless, as a foreigner, a Frenchman and a Catholic, Hubert was a perfect scapegoat. Ever maintaining his guilt, Hubert was brought to trial, found guilty and duly executed by hanging.

The extremes to which some Confessing Sams will go is illustrated by John Hart. In the 1920s, Hart confessed to the Jack the Ripper murders in late 19th-century London. Although it was pointed out that Hart was only 3 years old at the time of the first murder, this did not shake his story. He maintained he was Jack the Ripper for the rest of his life. More than 50 people confessed to having committed the famous and still unsolved Black Dahlia murder in Los Angeles in 1947. No one was ever charged. At least six people have confessed to being the Zodiac Killer who terrorised San Francisco on a murderous rampage beginning in 1968. One of the confessants is a woman — a 'Confessing Samantha'. The case is still unsolved. At last count, 20 individuals have confessed to the 1996 murder of child beauty queen Jon Benet Ramsey.

Why does someone become a 'Confessing Sam'? According to forensic psychiatrists, Drs Peter Quintieri and Kenneth Weiss,[19] there are many reasons why a person will give a non-coerced false confession. Among these are mental illness, mental retardation, attention-seeking, publicity-seeking, or a combination of these. Often a psychiatric disorder, one involving severe guilt feelings completely unrelated to the crime at hand, may provide the motivation.[20] Researchers ironically note that it is often just as difficult to know whether or not someone is telling the truth when they plead *against* themselves as when they plead *for* themselves. Dr S.M. Kassin and colleagues[21] report that when college students and police investigators judged 10 prison inmates confessing to crimes (half the confessions were true, half were false as they were concocted for the study), the students were more accurate than the police in determining who told the truth.[22, 23]

DOES SUBLIMINAL ADVERTISING WORK?
(Asked by Peter Ryan of Newcastle, New South Wales)

The term subliminal advertising was coined by market researcher James Vicary. It is the technique of exposing a person to products without the person being consciously aware of it. In theory, once exposed to the subliminal (subconscious) message, the person decodes the message and acts upon it while never knowing anything happened. In 1957, a bestselling book, *The Hidden Persuaders* by Vance Packard, warned of advertising camouflaging psychological 'subthreshold effect' to stimulate us to buy things against our conscious will. Vicary conducted an experiment to test the extent of this effect: Over a 6-week period, more than 45,000 people were exposed to a subliminal message flashed on a screen of a suburban New Jersey theatre that lasted for 3/1000 of a second. The message read 'Eat Popcorn and Drink Coca-Cola'. Vicary claimed that popcorn sales rose 57 per cent and Coca-Cola sales rose 18 per cent, though he never published his results. Nevertheless, a more than 50-year controversy was launched that continues today.

In books such as *Subliminal Seduction* (1974) and *The Age of Manipulation* (1992), Wilson Bryan Key has argued that subliminal messages permeate advertisements, especially sexually suggestive messages. All sorts of examples are presented. While some experts agree, most researchers have denied that subliminal advertising works. Deniers such as market psychologist T.E. Moore offer two chief reasons. First, subliminal messages are weak and are usually not even perceived. Second, even if perceived, people don't do what they consciously don't want to do. There is no such thing as a Manchurian Candidate Consumer.

Controversy continues. A fascinating recent experiment found that people who claim to have so-called ESP (extra-sensory perception) may also be more susceptible to subliminal messages.[24] In 2002, Dr Susan Crawley and colleagues[25] reported on their results of tests of people using Zener cards, which for years have been used to test for ESP. Each Zener card displays 1 of 5 symbols: A circle, a cross, a square, a star or 3 wavy lines. Dr Crawley's team had subjects sit in front of a computer monitor displaying the back of a Zener card. They pressed a key to choose which symbol they thought was on the face of the card. Then they got to see the card's face. What subjects didn't know was that they were also exposed to the correct answer flashed on the screen subliminally at 14.3 milliseconds before they made their choice. This is supposedly too fast for most people to perceive. However, it was found that people who claimed ESP abilities 'were able to subconsciously pick up on the clue, and as a result they scored better than chance at predicting which symbol would appear'.[26] This study lends support to both the existence of ESP and the power of subliminal messages to change behaviour.[27]

DO LIE DETECTOR TESTS REALLY WORK?
(Asked by Lisa Burnham of East London, UK)

A lie detector test or machine is a popular but inaccurate term for the instrument that records various bodily changes that may provide the basis for a reliable diagnosis of truth or falsehood. The correct term for the instrument is a polygraph. There is no such thing as a lie detector, lie detector test or lie detector machine if the terms are taken to mean a mechanical test or device that will produce a clear indication of lying when verbal statements are made. The polygraph technique, as it is more properly called, measures respiration, blood pressure and pulse. A supplemental unit that records the galvanic skin response (GSR) is also part of the procedure. The most valuable indicators are respiration and blood pressure and not, as is commonly believed, the sweating palms measured by the GSR. The polygraph technique is actually more of a diagnostic procedure than a mechanical operation. A high competence and skill level of the examiner and a high calibre of questions produce the most meaningful data. Polygraphs do not simply indicate whether a specific statement is true or false. A pattern of response compared to various control questions is evaluated by the examiner to determine the truthfulness of the answers. Results of the polygraph technique are almost never admissible in court as evidence unless both sides agree in advance to its use. However, it is still used in other places such as in the workplace to assess the honesty of employees. Whether the employees can make the boss take one is another matter.

◆

According to science writer Paul Marks, the US Department of Defense plans to develop a lie detector that can be used without the subject knowing they are being assessed.[28] The Remote Personnel Assessment (RPA) device will also be used to pinpoint fighters hiding in a combat zone, or even to spot signs of personal stress that might mark someone out as a terrorist or suicide bomber. The RPA will use microwaves or laser beams reflected off a subject's skin to assess various physiological signs without the need for wires or skin contact. The RPA will focus a beam on a moving or non-moving subject and use the reflected signal to calculate pulse, respiration or galvanic skin response.[29, 30]

HOW HONEST ARE HUMANS?

It is said that humans are as honest as the day is long. Unfortunately, the days are pretty short in winter. How honest are we? Three studies shed some light on aspects of this.

Are highly religious people more honest?

In 2006, Dr M.A. Huelsman and colleagues[31] reported on their findings from two standard behavioural surveys given to 70 of their students. One was the Santa Clara Strength of Religious Faith Questionnaire which measures degree of religiosity. The other was the Academic Practices Survey which measures things such as likelihood to practise plagiarism. The researchers found that 'overall, religiosity and academic dishonesty were not significantly related'. However, there was a slightly larger effect for women compared with men. So from this study at least, it would seem that religious people are not significantly more honest than non-religious people.[32]

Is most plagiarism intentional?

Perhaps it is not, at least according to some researchers. Plagiarism is the stealing or the passing off of the words or ideas of someone else without crediting the source. Supposedly, plagiarism is rife throughout our society. In 1993, Dr M.K. Johnson and colleagues[33] reported that people who plagiarise often forget where they have learned information. This makes them more likely to 'misattribute' it at a later date. Dr Johnson and colleagues point out that people have even been known to plagiarise themselves. Furthermore, they argue that research on non-conscious memory shows that people can continue using a skill even if they have forgotten that they learned it. So perhaps we should be a little less judgmental in at least some instances of plagiarism.[34] In any case, research suggests that plagiarism is much more difficult for students and others to get away with these days due to very sophisticated software packages that detect plagiarism.

Do adolescents value honesty more or less than they did in previous generations?

It seems that most adolescents are pretty honest these days — at least they value honesty rather robustly. In 2007, psychologists Drs S.A. Perkins and Elliot Turiel[35] reported on their study of 128 adolescents. The adolescents had to make a series of moral judgments in various hypothetical situations. The researchers were interested in finding out under what circumstances (if any) the adolescents would judge it moral to tell a lie. Drs Perkins and Turiel conclude that overall 'most adolescents thought that directives from parents and friends to engage in moral violations ... were not legitimate, whereas parental directives concerning prudential acts were seen as legitimate. Results indicate that adolescents value honesty, but sometimes subordinate it to moral and personal concerns in relationships of inequality.'[36] From these studies, perhaps the day just got a little longer.

◆

Unintentional plagiarism is known as cryptomnesia.[37]

WHY DO I SOMETIMES CHOOSE TO SUFFER EVEN WHEN I COULD AVOID IT?

Studies show that it is very easy to psychologically manipulate people to 'choose to suffer', such as by eating worms or enduring electric shocks, without much reason for doing so and even when they could avoid it.

In a classic experiment reported by Drs Robert Comer and James Laird in 1975, subjects were first led to expect that they would have to eat worms in order for a minor positive consequence to occur that had little bearing upon themselves (the experiment would be designated by officials as a 'success'). That was all it took for many subjects to eat the worms. For subjects who held out, they were told that there was no way they could escape eating worms without a minor negative consequence taking place that also had little bearing upon themselves (the experiment would be terminated). This brought more subjects around so that few now resisted eating the worms. But what happened next was surprising. After resigning themselves to eating worms and just before doing so, subjects were offered a choice between eating worms and performing a simple 'non-aversive' task as an alternative. Most subjects chose to eat the worms anyway! As an astonished Comer and Laird write, most subjects 'elected to suffer'.[38]

In another classic experiment reported by Dr Rebecca Curtis and colleagues[39] in 1984, subjects voluntarily administered electric shocks to themselves at varying voltages for the dubious gain of being told that they would become better test-takers in such experiments.[40] However, in yet another and more recent experiment in this odd and curious domain of behaviour, Dr Roy Baumeister and colleagues[41] published their results in 2002. Subjects risked their lives and health after merely being told that to do otherwise would result in their being 'likely to end up alone later in life'.[42] Strange what we do, is it not? For many, the price of suffering is cheap.

ARE THERE PEOPLE WHO CANNOT EXPERIENCE PLEASURE?
(Asked by Taylor Medved of Perth, Western Australia)

Apart from some people who are just plain grumpy, there is also a psychiatric condition known as anhedonia. With anhedonia there is the total absence of experiencing pleasure in the things that are normally pleasurable to everyone else. No matter what a person says, does or thinks, they do not receive any pleasure from it at all! Anhedonia literally means 'without pleasure'. *An* means 'without' and the Greek *hedone* means 'pleasure'. Anhedonia is sometimes seen in schizophrenics and epileptics, and is one of the earliest signs of a variety of neuroses. According to one theory of anhedonia, it occurs when the person's own psychological reward-value system is somehow blocked. They become emotionally numb to any pleasurable feelings at all. They are in for a pretty miserable life. It seems that all sorts of things can block the reward-value system. One of these is witnessing trauma, violence and the suffering of others. In 2006, Dr T.B. Kashdan and colleagues[43] reported on their study of US combat veterans now suffering from post-traumatic stress disorder (PTSD).[44] It seems that many with PTSD also suffer emotional numbing and (you guessed it) anhedonia. Woody Allen's 1977 film *Annie Hall* was originally entitled *Anhedonia*. In the film, Allen's character remarks that his belief is that life can be described as moving between the terrible and the horrible. Sounds like anhedonia alright![45]

WHAT IS ROMANTIC JEALOUSY?

(Asked by S. Johnson of New Rochelle, New York, USA)

Romantic jealousy is one of our strongest and strangest emotions. It can break hearts, tear couples apart, destroy lives and even cause depression and result in suicide. No matter how solid your relationship, it is vulnerable to feelings of romantic jealousy. We feel romantic jealousy when we perceive an external threat to a relationship that is important to us. The relationship's importance may be due to either an emotional or a sexual bond or the self-esteem or social prestige boost we get from it. When we become romantically jealous, a physical reaction commonly follows. The heart may pound, the knees may shake and the emotional flood can take the form of rage, fear, sadness, depression, or a combination of any of these. Romantic jealousy involves three people. Two have the relationship and the third threatens that relationship. We frequently feel justified in our romantic jealousy. We feel our territory has been invaded and we feel rightfully entitled to react. Charles Darwin in *The Descent of Man* (1871) claimed that romantic jealousy evolved to advance our species. Through romantic jealousy we ward off threats to our relationships and this helps ensure that our genes will be passed along to the next generation. Sigmund Freud theorised in his *Introductory Lectures on Psycho-analysis* (1922) that feelings of romantic jealousy first emerge during our early childhood from the Oedipus Complex (our desire not to be left out in the three-way mother–father–child relationship). If you find that you are romantically jealous, Dr Ayala Malach-Pines[46] suggests that you ask yourself the following: 1) What exactly am I feeling? 2) Are my feelings truly relevant and justified in the present situation? 3) What has triggered the romantic jealousy? Dr Pines also suggests that if you feel threatened in an existing relationship, you should discuss your emotions with your partner.[47]

WHY DO HUMANS SEEK REVENGE?
(Asked by Humphrey Johnson of Oklahoma City, Oklahoma, USA)

We humans do seem very revengeful, don't we? There are at least eight theories for why humans seek revenge. First, there is the 'defence function of revenge'. According to this theory, acts of revenge signal to a potential aggressor that if they act aggressively, the target of their aggression will defend themselves by striking back. Thus, the potential target person protects themselves against future attacks. Second, there is the 'pleasure function of revenge'. The famous psychologist Karen Horney (1885–1952) spoke of the 'vindictive triumph' — the feeling of excitement and elation — that can accompany the quest for and fulfilment of an act of revenge. Third, there is the 'restoration of pride function of revenge'. Fourth, there is the 'undoing the shame and humiliation generated by the original injury or loss function of revenge'. Fifth, there is the 'protecting from grief over a loss and anxiety resulting from separation or abandonment function of revenge'. Sixth, there is the 'avoidance or lessening of the painful aspects of mourning function of revenge'. Italian psychiatrist Franco Fornari (1921–1985) wrote of 'the paranoid elaboration of mourning' to describe this process of aggressive acting out as a substitute for doing the internal work of grieving. Seventh, there is the 'self-esteem building function of revenge'. As psychiatrist Dr Salman Akhtar[48] writes: 'Although typically viewed as politically incorrect, some revenge is actually good for the victims. It puts the victim's hitherto passive ego in an active position. This imparts a sense of mastery and enhances self-esteem.'[49] Eighth, there is the 'anger release function of revenge'. Psychiatrist Dr David Lotto[50] writes about the widespread US reaction of anger and revenge to the 9/11 disaster: 'Although there was a good deal of grief, sadness and feelings connected to loss and suffering expressed, what was most striking was the level of anger that emerged. The anger was largely contained and couched in the garb of rational intellectual discourse, but the power and intensity of the rage was palpable.'[51]

WHAT IS IMAGINED POVERTY SYNDROME?
(Asked by Ian Grant of London, UK)

Imagined Poverty Syndrome (IPS) is one of the strangest and least understood of all psychiatric conditions. It is not as yet recognised by the DSM of the American Psychiatric Association so it is not an official psychiatric condition, disorder or illness. Nevertheless, it is emerging as such. According to Dr Anthony N.G. Clark[52] who has studied 30 IPS cases, people with this mental illness dress in rags, live in filth and squalor, and may hoard vast piles of garbage — although they are economically comfortable, even wealthy. IPS sufferers are often of high intelligence, from excellent families, with fine education and have previously enjoyed good to great careers. Yet they needlessly live in poverty, not created by economic conditions, but by their own imagination. So little is studied about IPS that no one really understands why it occurs. In one of the few studies of IPS sufferers, it was found that many were born into wealth, others obtained it themselves, while half had been professionals. Doctors, lawyers, academics, officers in the armed forces, fashion designers and musicians were included.

The one common characteristic of all IPS sufferers is that they tended to be aloof, suspicious and hostile towards others. Few had any friends. Most had cut off all ties with all other people and lived in near or complete social isolation. All suffered from malnutrition and many had serious neglected diseases often because they were mistrustful of physicians and did not seek medical care. They were also extremely frugal. Dr Clark observes that 'it is painful for these people to spend money. Many believe they are paupers and end up ill from the lack of looking after themselves.'[53] But they are not paupers, not even close, as the following two cases illustrate:

> ⟡ A woman who dressed in rags and lived in a rundown house
> with the windows boarded up collapsed due to lack of food and
> was taken to the hospital. It was discovered she was worth over
> £300,000.
> ⟡ A retired teacher lived in a ground-level hovel and ate food
> scraps left on her windowsill by charitable neighbours. Yet it was
> discovered that she possessed 10 bank books with assets totalling
> over £80,000.

So strange we humans!

IN SOME PLACES IS IT NORMAL FOR PEOPLE TO HOLD ECONOMIC VALUES UNLIKE THOSE OF WESTERN CIVILISATION?
(Asked by Kelly O'Connor of Hartford, Connecticut, USA)

There are traditional cultures where most people do not share the Western economic values that we often take for granted. In one such example, Dr George Foster, the late anthropologist at the University of California at Berkeley, wrote of the 'image of limited good' among villagers he studied in Mexico. First in *A Primitive Mexican Economy*

(1942) and later in works such as *Empire's Children* and *The People of Tzintzuntzan* (1973), Foster wrote of Mexican villagers who believed that, quite the opposite of how we are led to think, all things that are good (wealth, health, good fortune, luck and happiness) are fixed and finite within the community. 'Good' is limited in quantity, hence the 'image of limited good'. Given this belief, all individuals are entitled to their fair share. If one individual has far more than their fair share, for whatever reason, this is viewed as immoral. Such a person would be regarded as selfish, an improper citizen and more or less a community vandal or thief. With the belief of the 'image of limited good', these Mexican villagers would therefore condemn as immoral many of our Western economic and business practices and social behaviours. Among these would be our allowance of the amassing of great fortunes while others are poor, the driving of business rivals into bankruptcy, the unwillingness of many to be charitable in heart as well as mind and the list goes on. Of course, modern Mexico and the impact of the global economy has swallowed up and transformed the traditional villages since the time of Foster's research. But the example of a different economic value system is real and stands to this day as an alternative to compare with our own economic ideology.[54]

CAN PHYSICAL APPEARANCE AFFECT MY EMPLOYMENT PROSPECTS?
(Asked by S. Burgess of One Tree Hill, South Australia)

Evidence suggests that physical appearance does indeed play a role in employment prospects. For decades research has demonstrated that cultural beliefs, social stereotypes and widespread biases and prejudices incline many people to make hiring decisions based on what you are rather than who you are. All other things being equal, if you as a job applicant are taller, thinner, of lighter skin and better dressed compared to another applicant then you are more likely to be hired. However, more subtle aspects of physical appearance also affect employment prospects. One of these is even the simple appearance factor of whether or not you wear a beard.

- In 1989, Dr W.E. Addison[55] showed that bearded men 'were rated significantly higher on masculinity, aggressiveness, dominance and strength'.[56]
- In 2003, Drs Michael Shannon and C. Patrick Stark[57] had subjects evaluate nine equally qualified men for management trainee positions based solely on a photograph. They found 'that bearded applicants, although evaluated equally with non-bearded applicants, were selected for management positions at lower rates'.[58]
- In 2003, a Brazilian study by Dr A.A. de Souza and colleagues[59] found that 50 personnel managers (28 men and 22 women) who made hiring decisions at different companies in Sao Paulo 'clearly preferred clean shaven over bearded, moustached, or goateed men as prospective employees'.[60]

So, although Dr Addison found that beards signal greater masculinity, it may not be enough to help land the job. And if Harry is hairy, he may be harried.

◆

Phgonophobia is the morbid fear of beards.[61]

HOW POWERFUL IS REJECTION?
(Asked by B. Jones of Toronto, Ontario, Canada)

Research suggests that rejection has a powerful impact on us all. It doesn't just make you feel bad. Rejection lowers your IQ, makes you perform worse, increases your aggressive behaviour towards others, lowers your ability to accurately measure things such as time intervals, makes you more likely to commit judgment errors in the social realm and hinders your ability to act with self-control. Dr Roy Baumeister and colleagues[62] have presented several studies over the last decade or so highlighting the effects on us of being rejected by others. In one study published in 2002,[63] Dr Baumeister and colleagues describe one of their experiments and what they concluded: subjects were given a variety of intelligence and analytical skills tests. They were then made to feel rejected. Some were given a personality evaluation that led them to believe (falsely) that they were destined to spend their lives alone. Others were allowed to mingle with a group of strangers with whom the subjects were told they would soon be called on to complete a task together. But later the subjects were told that none of the strangers wished to have anything to do with them. After these experiences of rejection, the subjects were tested again for intelligence. To the researchers' surprise, the IQ scores of subjects plummeted by some 25 per cent on average and their analytical reasoning skills declined by about 30 per cent on average. Baumeister and colleagues offer an explanation: 'Connecting with others is one of the deepest and most powerful human drives we have. Thwarting it has a big impact on us. After being rejected, people cannot think straight for a while.' When people reject us, we act dumber than we really are. Rejection really hurts us in more ways than one.[64]

IS PARKINSON'S DISEASE RELATED TO PARKINSON'S LAW?
(Asked by Jackie Stokes of Detroit, Michigan, USA)

Parkinson's disease and Parkinson's Law are completely unrelated. Parkinson's disease is also known as Parkinson's, Parkinson's syndrome, Parkinsonism, *paralysis agitans* and shaking palsy. By whatever name, it is a degenerative disorder of the central nervous system that is characterised by tremor and impaired muscular coordination.

Parkinson's Law has nothing to do with illness or with the body for that matter. But it does have something to do with behaviour.

Parkinson's Law states that 'work expands so as to fill the time available for its completion'. This implies that those performing the job multiply at a fixed rate, regardless of the amount of work produced. Parkinson's Law was formulated by C. Northcote Parkinson (1909–1993), a British political scientist. He poked fun at modern bureaucracies in his 1957 book, *Parkinson's Law: the Pursuit of Progress*. Parkinson's classic illustration of the law that bears his name is the contrasting of the amount of time taken to write a letter by a busy businessman and by his maiden aunt: any job will take the amount of time available. Parkinson made a number of other clever observations about modern life. Among these are 'Expenditure rises to meet income'; 'Time spent on any item of the agenda will be in inverse proportion to the sum involved'; 'The man who is denied the opportunity of taking decisions of importance begins to regard as important the decisions he is allowed to take'; and 'Men enter local politics solely as a result of being unhappily married'.

A corollary is *Hofstadter's Law*, devised by Douglas Hofstadter, a US computer scientist. Hofstadter's Law is in two parts. First, any given job always takes longer than you expect. Second, any given job always takes longer than you expect even when you take Hofstadter's Law into account. Parkinson's Law is often confused with The Peter Principle. *The Peter Principle* states that in a hierarchy, every employee tends to rise to their level of incompetence. In other words, as long as you do your job well you're likely to be promoted. Only when you do not do your job well will your promotion be stopped. Thus, everyone finishes their promotion curve at the point of their incompetence — and every worker who has been at a job for a long time is bound to be incompetent at it. This has a devastating effect on the efficiency of businesses, bureaucracies and organisations. This theory was formulated by Dr Lawrence J. Peter while a professor of education at the University of Southern California. A corollary to The Peter Principle is *The Dilbert Principle* which is named after the comic strip character and coined by Scott Adams: Companies tend to systematically promote their least competent employees to management.[65]

CAN A PERSON BE ADDICTED TO THE INTERNET?
(Asked by Ian Anderson of Aberdeen, Scotland)

Internet Addiction Disorder (IAD) is one of the new psychopathologies of the internet era. The first mention of IAD was in a 1996 paper by Drs O. Egger and M. Rauterberg.[66, 67] The first case of IAD in the clinical literature was presented by Dr K.S. Young.[68, 69] The case concerns a 43-year-old housewife who was addicted to the internet yet who otherwise had no prior history of any other psychiatric problem. It is unknown how many people suffer from IAD. There are several symptoms of IAD. These include:

- A need for an ever-increasing amount of time on the internet to achieve satisfaction or a dissatisfaction with the continued use of the same amount of time on the internet.
- Two or more withdrawal symptoms developing within days, weeks, or up to a month after a reduction or cessation of internet use. These include distress or impairment of social, personal or occupational functioning such that there is psychological or psychomotor agitation such as anxiety, restlessness, irritability, trembling, tremors, voluntary or involuntary typing movements of the fingers, obsessive thinking, fantasies or dreams about the internet.
- Internet engagement to relieve or avoid withdrawal symptoms.
- Internet often accessed more often or for longer periods of time than was intended.
- A significant amount of time is spent in activities related to internet use, for example, internet surfing.

• Important social, occupational or recreational activities eliminated or reduced due to internet use.

• Risk of loss of a significant relationship, job, educational or career opportunity due to excessive internet use.

• Internet engagement used as a way of escaping problems or relieving feelings of guilt, helplessness, anxiety or depression.

• Concealing from or lying to family members about the extent of internet use.

• Internet user driven to financial difficulty due to incurring unaffordable internet fees.[70]

◆

Internet addicts typically spend 10 hours of internet 'recreation' for every 1 hour of internet 'work'.

◆

A 28-year-old Korean man died of dehydration in 2005 after spending 50 hours non-stop at an internet café.

◆

The Beijing Military Clinic in China is the first hospital devoted to internet addiction.

WHY DO I ENJOY GAMBLING SO MUCH?

(Asked by Susan Quarry of Norwich, East Anglia, UK)

All kinds of drug and non-drug experiences, including gambling, exercise, computer use, video games and even love, could be described as addictions. This was the stunning theory put forward by social psychologist Dr Stanton Peele first in *Love and Addiction* (1975) and more recently in *7 Tools to Beat Addiction* (2004). A 2000 survey commissioned by the British National Centre for Social Research revealed that about 1 per cent of the UK adult population had a pathological gambling problem, especially women. Some authorities believe that up to 10 per cent of the US population has a gambling problem. It is estimated that there are 23 million internet gamblers of whom 4 million exist in the UK and 8 million exist in the US. There are now more than 2000 online gambling sites with an estimated global internet gambling revenue in 2005 of $82.2 billion. This is expected to rise by 50 per cent over the next 5 years. Surveys show that 75 per cent of US adults gambled at some time in the past year.

The nature of addiction is very controversial in both behavioural and medical sciences. It seems that excessive drug use or behaviours somehow trigger a rise in dopamine release in the brain within the 'reward circuitry' of the brain. Dopamine is a chemical naturally produced in the body. It is both a neurotransmitter and a neurohormone and the release of dopamine makes you feel very good. The dopamine release system involves four areas of the brain: the nucleus accumbens, the amygdale, the substantia nigra and the ventral tegmental area. The sympathetic nervous system and the central nervous system are affected. Simply put, addictive behaviours become addictive and stay that way because you feel so good when you do them.[71]

CAN YOU BECOME INTOXICATED MERELY BY THE POWER OF SUGGESTION?

(Asked by Jeff Johnson of Sheffield, UK)

Mind over matter goes a long way. Reports of intoxication occurring merely through the power of suggestion and not through alcohol ingestion appear from time to time. An early example of this is a 19th-century incident in the US state of Maine. A logging camp was stocked with bottles of vanilla extract containing alcohol. Workers at the camp would occasionally break into the camp's stores, drink the vanilla extract and become intoxicated. Eventually the logging camp managers changed to stocking bottles of vanilla extract not containing alcohol. The workers still occasionally broke into the stores, still drank the vanilla extract and *still* got intoxicated — without alcohol! Mind over matter can happen with non-alcoholic drinks too. 'What you think may be as important as what you drink.' This is according to Dr Andrew Scholey,[72] who presented a paper to the British Psychological Society in 2000 on the psychological effects on drinkers of caffeinated and decaffeinated coffee.[73]

As everyone is told, caffeinated coffee (CC) makes you stay awake and keeps you more alert; decaffeinated coffee (DC) supposedly does not. Dr Scholey and colleagues conducted a simple experiment. They informed each subject in the experiment, all of whom were coffee drinkers, that each would be assigned to one of two groups, A and B. They informed subjects that those in A would be given CC and those in B would be given DC. Each subject would then take a computerised test and would be told which group they were in (and what they were given). In reality, without being informed, all subjects were divided into four groups — A1, A2, B1, B2. Those in A1 were told they were getting CC and were given CC. Those in A2 were told they were getting CC but were given DC. Those in B1 were told they were getting DC and were given DC. Those in B2 were told they were getting DC and were given CC. The researchers found that, as predicted, subjects who drank CC were faster and more accurate on a computerised test — but only if they thought they had been given CC. Subjects who drank CC but thought it was DC performed less well. Most interesting of all, subjects who thought they had drunk CC, but in reality had drunk DC, performed on the tests as if they had really drunk CC. So what you *think* is real can be more important than what *is* real. Mind over matter goes a long way.[74, 75]

WHAT IS DENIAL AND WHY ARE PEOPLE SO OFTEN IN DENIAL?
(Asked by Ron James of Manchester, UK)

In the psychological sense, denial is a defence mechanism in which a person, faced with a painful fact, rejects the reality of that fact. They will insist that the fact is not true despite what may be overwhelming and irrefutable evidence. There are three forms of denial. *Simple denial* is when the painful fact is denied altogether. *Minimisational denial* is when the painful fact is admitted but its seriousness is downplayed. *Transference denial* is when the painful fact is admitted, the seriousness also admitted, but one's moral responsibility in the situation involving the painful fact is downplayed. When a person is in denial, they engage in *distractive* or *escapist* strategies to reduce stress and help them cope. The effect on psychological wellbeing in doing this is unclear.

The concept of denial was formulated by Sigmund Freud and greatly elaborated on by his daughter Anna Freud (1895–1982) in the second volume (1936) of her eight volume *Writings of Anna Freud*. The concept has been around for many decades. Denial is an important factor in public health. Drs M.S. Vos and J.C. de Haes[76] point out that, based on their study of cancer patients, up to 47 per cent of patients deny the fact that they have been diagnosed with cancer; up to 70 per cent deny the impact of the diagnosis upon their lives; and up to 42 per cent deny that it has any effect upon their feelings. They add: 'From a psychoanalytical viewpoint, denial is a pathological, ineffective defence mechanism ... On the other hand, according to the stress and coping model, denial can be seen as an adaptive strategy to protect against overwhelming events and feelings.'[77] Therein lies the appeal of denial to humans.

Denial allows someone to keep going unchanged despite reality. Denial is the path of psychological and moral least resistance. Five years after the 9/11 attack on New York and Washington DC, 40 per cent of New Yorkers were still to varying degrees 'fearful', 'traumatised' or otherwise 'unable to face reality' according to New York public mental health experts. In such a psychological state, people are not at their reasoning best — easily confused, manipulated and fooled. While in denial about global warming, people don't have to think about anything, inform themselves, change their consumption patterns, become actively involved in reforms or alter their behaviour in any way. Politicians with transference denial can absolve themselves of any moral imperative to take the necessary policy initiatives that scientists say are mandatory for our species to survive.[78]

PERCEPTICIDE: DELUDING OURSELVES INTO BELIEVING 'I SAW NOTHING, I KNEW NOTHING'

It is a sad truth of human behaviour that when atrocities are committed against a political, ethnic or religious minority, often the majority in that nation refuses to believe that anything unusual is happening. Behavioural scientists observe that such widespread denial is the first coping strategy used during such times of death and terror. This is followed by rationalisations to explain away the horrors. It is a behavioural phenomenon known as 'percepticide' — the killing of one's psychological ability to recognise, admit and then humanely respond to reality. Perhaps the most prominent example of percepticide over the last century or so is the German people's often-stated reaction to the Nazi holocaust: 'I saw nothing, I knew nothing.' But there is no shortage of other percepticide instances: Greece during the Colonels, Cambodia during the Khmer Rouge, Uganda during Idi Amin, Romania during Ceausescu, the Rwanda genocide, the ethnic cleansing in the former Yugoslavia and sadly several others.

Percepticide was coined by Argentine psychiatrist Dr Juan Carlos Kusnetzoff in a 1985 article in the Argentine journal *Revista de Psicoanalis*. Ironically, Argentina during the 'Dirty War' (1976–1983) provides a clear instance of percepticide in action. Dr Marcelo Suarez-Orozco[79] writes that 'there is widespread evidence that during 1976 and 1977, at a time when hundreds of people were disappearing on a daily basis, Argentines largely refused to believe the extent of the atrocities committed in their country'. At the peak of the terror, most individuals coped 'by practicing denial and rationalisation on a large scale'. Indeed, 'the perceptual organs, too, soon became a casualty of the engulfing terror'. Percepticide emerged as 'the major coping response to terror'. A repressive political regime murders, tortures, imprisons without trial, or causes its opponents to 'disappear' often in the name of 'national security'. Ironically, according to Suarez-Orozco, percepticide exists because the individual needs psychological 'internal security'. In fact, such a need is greater than maintaining a clear view of outside reality. Thus, denial is allowed to block perception when what would be perceived is simply too terrible to accept. Suarez-Orozco maintains that in percepticide there is 'a failure to recognise certain aspects of the environment and the self' and the 'ego's reality-testing capacities are affected'. He cites a typical case of percepticide. After being informed that her siblings had 'disappeared' [were desaperecidos] in the midst of the terror, a young woman convinced herself that not only had her own siblings not 'disappeared', but also that there were no desaperecidos at all in Argentina. Countless similar cases repeat this pattern. In brief, Argentines developed a 'passion for ignorance'.

Suarez-Orozco notes percepticide was also promoted by the lack of bodies during the disappearances. He claims that this 'subverted the mourning process and added fuel to the tendency to deny: without a corpse to ritually mourn, there is always the fantasy that the person is not really dead, the atrocities not real'. It is very important to a regime committing atrocities that the population not see evidence of those atrocities. So news coverage of funerals are banned, news censored and commentary tightly controlled. Nevertheless, a percepticidal denial can only serve so long. Suarez-Orozco observes that in Argentina denial eventually gave way to two percepticidal rationalisations. In the first rationalisation, if the individual's initial response to the unspeakable was 'No it cannot be happening', in the face of continued information and rumours to the effect that it is happening, then the second response was 'It cannot happen to me or my family, it can only happen to those involved in something bad'. In fact, the Argentines have a phrase for this, *estaria metido en algo* [He/she must have been involved in something subversive]. However, use of this first rationalisation becomes shaky indeed when used to explain away a *desaparecido* known to the person not to be subversive or otherwise obviously innocent. In the second rationalisation, when it was revealed and widely reported that dozens of babies who were born to women political prisoners were themselves also made to 'disappear', Suarez-Orozco writes that 'an alternative modal rationalisation' was invoked: that 'the whole thing was part of an elaborate, invented, left-wing propaganda campaign' invented by enemies of Argentina. Indeed, 'a significant number of Argentines were at one point convinced that the reports of gross human rights violations that flourished in the international press were simply part of an infamous "anti-Argentine" campaign orchestrated by trouble-makers who had escaped to Europe and Mexico to "discredit" the fatherland. There is an inescapable "blame-the-victim" quality to such rationalisations. The atrocities cannot really be happening; they must be fabricated by the supposed victims who are enjoying refugee status in Mexico and Europe.' As with physical survival, humans will go to great extremes to psychologically survive. Denying the undeniable is one such extreme — deluding ourselves into believing 'I saw nothing, I knew nothing'.[80]

ARE SOME PEOPLE REALLY ACCIDENT-PRONE?
(Asked by Charles Haywood of Cedar Rapids, Iowa, USA)

Accident-prone means that one suffers a greater number of accidents than normal. Researchers are trying to discover if there is a certain type of person who is accident-prone. A few studies reveal a few clues. A French team of public health researchers, led by Dr G.C. Gauchard[81] attempted to identify the determinants of accident-proneness. They studied 2610 French railway workers and reported their findings in

2006.[82] The Gauchard team found that 27 per cent of the individuals they studied had more frequent than usual accidents with injuries. This was much higher than the researchers suspected. The researchers also found that youth, inexperience on the job, dissatisfaction with the job (indicated by applying for a job transfer), having no safety training, having a sleep disorder, smoking and getting little or no exercise were all related to suffering more accidental injuries. Surprisingly, there was another factor too: not having a personal hobby (such as gardening). According to the BBC, in 2001 a team of British researchers led by now emeritus Professor Ivan Robertson[83] identified three key personality traits of people who are *not* accident-prone:

1. *Openness*. This is the tendency to learn from experience and to be open to suggestions from others. But the Robertson team cautions that too much openness can increase accident risk.

2. *Dependability*. This is the tendency to be conscientious and socially responsible.

3. *Agreeableness*. This is the tendency not to be aggressive or self-centred. The Robertson team argues that people with low levels of agreeableness tend to be highly competitive and less likely to, for example, comply with safety instructions.[84]

❖

When it comes to accidents, some people seem to be truly star-crossed. Take the sad case of Thomas L. Cook as reported by the *Denver Post* in 2006. Cook got off to a poor start in life and it never got any better. Cook's accident-proneness started before he was born. He almost died before birth as his mother nearly miscarried. As a child he suffered many serious accidents. He broke his collarbone, suffered brain haemorrhage due to a playground accident and had his spleen removed due to an injury playing touch football. He then had a go-cart accident while a teenager, a near-fatal car accident before attending university and spent 5 months in a coma due to another car accident while at university. While employed as a computer programmer, Cook broke his back three times and broke ribs in various car accidents and falls. To his credit, he fought back from serious injury to regain his health. As Claire Martin writes in her *Denver Post* story: 'Thomas L. Cook, who died at 54 when he was fatally hit by a car September 11, spent much of his life recovering from the misadventures that plagued him even in the womb.'

Sometimes life is just not fair.[85, 86]

◆

Actor Orlando Bloom is notable for his accident-proneness. So far he has suffered fractures to the nose, back, skull (three times), right leg, left leg, a finger, a toe, a rib, a wrist and an arm.

ARE SECRETIVE PEOPLE MORE HEALTHY OR LESS HEALTHY?
(Asked by Karen Landon of Fort Lauderdale, Florida, USA)

Psychological research shows that keeping secrets, especially distressing ones, can make the secret keeper sick. But we do not know if keeping a secret per se causes more illness symptoms or if it is something about the type of person who is secretive that tends to make them sicker. For a century, psychiatrists, psychologists and other clinicians have noted that patients often hold back important information from their therapist. This is even if they want to get better and they know the therapist wants them to get better too. This is called 'self-concealment'. The self-concealer keeps secrets that are perhaps too painful to recall, too stressful to reveal, or even too frightening to describe. In 'On Beginning the Treatment' (1913), Sigmund Freud described the physical and psychological consequences of patients concealing information from the analyst. It is now known that patients self-conceal in both long-term and short-term therapeutic situations. And they keep both large secrets as well as small ones. Secrets seem to be of all kinds too. Judging from what patients do, more people are self-concealers than not. So it appears we humans are a very secretive lot — even with our own therapist.

Researchers have only recently begun to investigate how secrecy and nondisclosure can influence the therapeutic process and affect the patient's physical and mental health. In 1990, psychologists Drs Dale Larson and R.L. Chastain[87] reported evidence showing that 'people who are high in self-concealment (that is, the predisposition to keep secrets), as compared to those who are low in self-concealment, report having more physical and psychological symptoms'.[88] In 1997, Dr J.W. Pennebaker[89] proposed a three-part theory for why secrecy seems to be harmful. First, keeping a secret takes effort. Such inhibiting of behaviour requires the brain and body to work harder than it otherwise would. Second, when people cannot or otherwise do not reveal themselves, there is an increased probability of having obsessive thoughts developing around the secret. This also requires the brain and body to work harder. Third, conversely, the act of revealing 'reduces autonomic activity'. So, the theory goes, telling a secret lifts a burden from mind, heart and soul. Life is easier and the patient becomes healthier.[90]

In 2000, Dr C.E. Hill and colleagues[91] observed that 'secrecy may require considerable psychic energy which can leave the person with less energy to deal with other important issues'. They also argued that secretive people 'come to feel inauthentic, fearing that they are accepted only for the social mask they wear for their therapist'. This too causes stress and stress causes illness.[92] But what of the question, is it keeping a secret per se or something about the type of person who is secretive that tends to make people sicker? In 2006, Drs Anita Kelly and Jonathan Yip[93] presented evidence showing that the process of keeping a secret predicts fewer symptoms, whereas the personality trait of 'high self-concealment' predicted more symptoms. They put 86 volunteers through a battery of tests and arrived at this conclusion. The next step for researchers is to tease out what is the personality trait of secretive people that tends to make them sicker.[94] In any case, Grandma was right. Make a clean breast of it to your therapist. It is good for you — and that's no secret.[95]

CAN ABSENCE REALLY MAKE THE HEART GROW FONDER?
(Asked by Pat Cole of North Curl Curl, New South Wales)

Psychologists now recognise that there is a real challenge of so-called 'long distance relationships'. These relationships test the notion that absence makes the heart grow fonder and also test the budget limits for long distance phone calls and travel. Typically, 'long distance relationships' involve two individuals who work in different locations that make seeing each other very often impossible. Do such relationships get stronger? According to Dr Marie Hartwell-Walker,[96] long distance relationships can work if a few rules are followed:

- The couple is committed to the commitment.
- Both keep their partners visible to the people around them as well as to themselves.
- The arrangement meets the needs of both.
- The arrangement is within each partner's physical 'intimacy zone'.
- Both are focused on their careers when they are working and on each other when they are together.
- They consider carefully whether they have what it takes to add a 'third career' (child-rearing) to the mix.

Absence *can* make the heart grow fonder, but so too *can* familiarity breed contempt.[97]

WHICH COMES FIRST, IMAGINATION OR FANTASY, AND WHAT IS THE DIFFERENCE BETWEEN THE TWO?
(Asked by Mike Valentine of Fort Mill, South Carolina, USA)

The word imagination comes from the Latin word *imaginare* and means 'to form an image or to represent'. The imagination is the synthesis of mental images into new ideas. It is the power of the mind to form mental representations of a thought, concept, dream, symbol or fantasy. The imagination has the power to create or re-create any sensation perceived or possibly perceived by the mind. Since sensation is reliant upon the body and the brain, the imagination cannot be entirely separated from either. Imagination is not the opposite of reality. Instead, it is a means of adapting to reality. As such, imagination is essential to reality. Mental life could not exist without the imagination. Certainly creativity would be impossible without the imagination. The imagination is one of the most unique and important aspects of being a human being.

The word fantasy comes from the Greek *phantazesthai* and means 'picture to oneself' — and that is exactly what it is. A fantasy is a product of the imagination. It is an imagined sequence of events or mental images, for example, a daydream. It is a form of a story from the self with one common factor: the subject imagining the fantasy appears as one of the actors in the story. A fantasy may originate from conflicts, desires, frustrations or wishes when the imagination interacts with reality. A fantasy may substitute for action or pave the way for later action. In Freudian terms, the fantasy may itself afford gratification for id impulses (our darkest and most hidden drives), may serve the ego as a defence (our recognised self), or may take over super-ego functions by providing the imagery on which concepts are based (our self we project to others). A fantasy can be a conscious or unconscious construction. When it is unconscious, it is sometimes spelled 'phantasy'. Again, mental life could not exist without fantasy. And fantasy is also a unique and important aspect of being a human being.[98, 99]

❖

An article in the *Journal of Personality and Social Psychology* (December 1999) concludes, among other things, that people who have difficulty recognising their own incompetence often have 'inflated self-assessments'.[100, 101]

❖

The year 1953 was the last year in which no film, screenplay or performance relating to mental illness was nominated for an Oscar.

Chapter 14
Endings

(Asked by Jackie Bainbridge of Brooklyn, New York, USA)

There is probably no limit to the number of times a person can almost die and be brought back to life. But you would not want to test the principle either. There is a fascinating condition called Reflex Anoxic Seizure (RAS) where a person *seems* to clinically die and then come back to life many many times. In one case, a New Zealand girl 'died' and recovered 35 times between 14 months and 36 months of age.[1] RAS is a rare and incurable condition. The slightest bump or fall sets into action a frightening chain of events. First, the lungs collapse and the heart stops beating. Then the person suddenly begins to writhe and jerk wildly as the body is overcome by a major seizure. Finally the muscles relax and the person plunges into a death-like, catatonic, comatose state, which can last for hours. The person could easily be mistaken for really being dead if doctors don't know they have RAS. It can even stimulate taphephobia — the morbid fear of being buried alive.

According to Dr J.B. Stephenson,[2] who has been writing about RAS since 1978,[3] the condition should not be confused with an epileptic seizure. In RAS, the seizure involves a faint (syncope) being the trigger that is accompanied by odd, so-called 'white breath-holding attacks' called 'cardioinhibition'. Very strange indeed![4]

IN SOME PLACES IS IT NORMAL FOR PEOPLE TO LIVE TO BE 120 OR MORE?

(Asked by Kelly O'Connor of Hartford, Connecticut, USA)

It is a myth that it is common to live to be 120 or more in some parts of the world. The facts do not support stories of communities of legendary centenarians. The three principal candidates for such long-lived status are the Abkhasians of Georgia, the Hunzas of Pakistan and the Vilcambans of Ecuador. But other peoples from Tibet, India, China and elsewhere are sometimes suggested as extremely long-lived. Recipes for their longevity usually include being members of a small population from an isolated region, living an outdoor existence, following a simple subsistence diet (usually high in vegetables and fruit but low in meat), having a lean body mass, holding a cultural disapproval of excess body fat, being physically active (especially by walking), working in a job that is not too strenuous, following a stress-free life, not smoking tobacco (but alcohol consumption is OK), and having a strong cultural respect for old age. Yet certain curious factors are evident in these reports of longevity. There is usually an absence in these groups of a proportionate number of elders between the ages of 70 and 90, there are no birth records, the people are illiterate, age is often exaggerated in accounts of folk history, and there is a poor regard for the accuracy of time generally. In the case of the Abkhasians, years have likely been added to their ages in the past in order to avoid military service. No person with a verifiable record has ever exceeded 110 years by more than a few years, let alone most or even many of a whole community surpassing it by 10 to 40 years.

◆

An Abkhasian story holds that one man who insisted he was 'only 95' wanted to marry again. He became angry when it was pointed out to his bride-to-be that he must be older than 95 since he had a daughter who was 81. It was claimed by others that he was really 108.[5, 6, 7]

◆

Some cell biologists now believe that there is a definite biological limit to human cell reproduction making the maximum age possible for human life to be about 110 to 120 years. However, others maintain that humans could theoretically live to be 400 or 500 years old once genetically engineered drugs are developed to counteract the ageing process itself on the cellular level.

◆

The death of a husband does not appear to affect the longevity of his widow. But the death of a wife shortens the longevity of her widower.

WHAT IS THE MADAME BUTTERFLY EFFECT?
(Asked by Gina Hadley of Mt Vernon, New York, USA)

There is some evidence that opera lovers are more prone to suicide than non-opera lovers. This has been called the Madame Butterfly Effect. Researchers are not sure why opera lovers as a group are more suicide-prone. As Dr Steven Stack[8] theorised in 2002, suicide figures as a theme in opera quite frequently, especially the notion of suicide as a choice in the face of loss of honour. This occurs famously in Madame Butterfly. Dr Stack speculates that opera lovers may hold a higher sense of honour than non-opera lovers. In any case, he has found that in a survey of 845 subjects, 'opera fans are 2.37 times more accepting of suicide because of dishonour than non-fans'.[9]

WHAT CAUSES DEATH WHEN SOMEONE IS CRUCIFIED?

Most people think it is exposure, dehydration or even blood loss. But what surprises many is that the correct answer is suffocation. Crucifixion is a particularly brutal and painful form of execution. The ancient Assyrians and Babylonians invented crucifixion in the

6th century BC. The Romans perfected it over 500 years.[10] They usually applied it to criminals thought to be the lowest of the low: runaway slaves, disgraced soldiers, foreigners and early Christians among others. It was abolished by the Emperor Constantine in the 4th century AD.

A cross was laid flat on the ground and the victim was held down onto it. The arms were nailed into the ends of the horizontal plank through the wrist so that the carpal bones would prevent the nail being torn away through the hand. The legs were bent and twisted so that the feet were held sideways against the upright plank. A nail went through the heels on both feet behind the Achilles tendon. The cross was then hoisted upright into a prepared hole. The victim was then left hanging by the nails through the wrists. This position compresses the lungs making it very difficult to breathe. Death comes through slow suffocation as the legs lose their strength to support the body. The hapless victim attempts to change position to alleviate their suffering, but eventually their muscles cannot hold their weight and start to spasm. Death can take from 6 hours to 4 days.

Death can be sooner or later depending on whether or not the victim is tortured, scourged, maimed or otherwise injured before or during the crucifixion. Haemorrhage and dehydration can cause the blood vessels to break down (hypovolaemic shock) and the blood cannot carry enough oxygen (anoxaemia). Cardiac arrest eventually takes place. Roman guards could only leave the execution site after the victim died. They sometimes hurried matters along by fracturing the victim's leg bones, delivering a sharp blow to the front of the chest, or stabbing them with a spear. Drs F.P. Retief and L. Cilliers[11] report that in Roman crucifixions a fire was sometimes built at the foot of the cross. Wet straw was added to produce so much smoke that the victim would die of smoke inhalation.[12]

WHAT HAPPENS WHEN YOU'RE HIT BY A TASER OR STUN GUN?
(Asked by Leanne Foster of Potts Point, New South Wales)

Tasers and stun guns are high-voltage, low-current stimulators that can cause involuntary muscle contractions, loss of body control and sensations such as pain and extreme fatigue. They are used as police weapons since a person can be immobilised without the injury or death that would normally result from the use of a normal police revolver. Both tasers and stun guns produce electrical stimuli in the form of shocks of about 50,000 volts that last for a fraction of a second. In a taser, an electrode is shot out as a dart and impacts upon the body. In a stun gun, the electrodes are fixed into the gun itself. If taser darts are shot into a standing person's thigh at 25 cm (10 inches) apart and from a distance of 1.8 m (6 feet) away, the leg will be locked into a flexed position and the person will be unable to continue standing. If a stun gun is shot at a standing person's rib cage from a distance of 1.8 m (6 feet) away for 4 to 5 seconds, the person will be considerably weakened and in most cases brought to their knees.

There are surprisingly few studies in the scientific literature as to the effects on the body of tasers and stun guns. One of the few early ones appearing was by bioengineers Drs Raymond Fish and Leslie Geddes.[13] They suggest that research shows that tasers and stun guns may not be as safe as many of us are led to believe. Instead, tasers and stun guns can cause cardiac arrest and respiratory failure, disrupt pace-makers, damage eyes, injure the central nervous system and cause death. Death can result, in particular, if someone has taken certain drugs — something that is often the case when arrests take place. It is quite possible that tasers and stun guns may result in more deaths, not fewer, as police are more likely to use them thinking they are safer. So the two doctors argue that tasers and stun guns should not be used, for example, to restrain drunk drivers. They can also cause miscarriages when used against pregnant women. Police would find it difficult, if not impossible, to know if a female suspect is pregnant.[14]

WHAT IS YOUR CHANCE OF BEING SHOT AND KILLED BY A TASER OR STUN GUN?

(Asked by Ian Leonard of London, UK)

According to one study, the chance of dying after being shot by a taser or stun gun is about 1 in 870. This is the estimate of Dr William P. Bozeman.[15] But Dr Bozeman acknowledges that any such estimate is based on very little data. Tasers and stun guns have been in use for only a short period of time and there are still few studies assessing their various effects.[16]

- Since 2001, medical and legal studies have suggested that more not fewer deaths could result from the introduction of tasers and stun guns in police work. Some doctors and civil libertarians worry that more deaths could occur since police may be more likely to use tasers and stun guns on a suspect thinking they are safer than handguns.
- Tasers and stun guns were first introduced in police work in South Africa in 1999.
- Of 75 people who have died after being shot with a taser or stun gun, the taser or stun gun was considered a potentially contributory cause of death in 27 per cent of cases.[17]
- The use of cocaine lessens the shock value of a taser or stun gun by 50 to 100 per cent.[18]
- Police perform a very important and very difficult job. Use of tasers and stun guns results in fewer injuries to police compared with using clubs, dogs, handguns or nothing at all to restrain suspects.[19, 20]

DO CHILDREN OF ATOMIC BOMB SURVIVORS DIE EARLIER?
(Asked by Jim Cranston of Atlanta, Georgia, USA)

After the atomic bombings of the Japanese cities of Hiroshima and Nagasaki at the end of World War II in August 1945, it was feared that the radiation damage done to the survivors who could still have children would be passed onto their offspring. It was even thought that such children would die rapidly, suffer great deformities and pass on genetic mutations to their children. The latest evidence indicates that these fears may be unwarranted — at least in one sense: life expectancy. Research shows that children born to survivors of the World War II atomic bombings do *not* show any evidence of dying earlier than normal. However, the researchers caution that these individuals may still die earlier, but they haven't done so as yet. Japan has the longest life expectancy in the world at 84.6 years for females and 77.9 years for males. So it will be a decade or so before we know for sure.

Dr S. Izumi and colleagues[21] published their latest review of the mortality statistics among 'offspring of atomic bomb survivors' in 2003. In all, 41,010 subjects are being studied. All were born between May 1946 and December 1984. Each subject has at least one parent who survived either the Hiroshima or the Nagasaki bombing. The researchers observe that cancer and non-cancer mortality rates were no higher for subjects with exposed parents. Overall, few have died — the same for this age bracket elsewhere in Japan. It was also found that it did not matter whether or not both parents were 'atomic survivors' or just one. And the dosage level of radiation exposure (high, low, or somewhere in between) made no difference to mortality either.[22] Cross your fingers for the future!

DO CHILDREN OF US ATOMIC VETERANS DIE EARLIER?
(Asked by Jim Cranston of Atlanta, Georgia, USA)

The US Department of Veterans Affairs (DVA) classifies 405,000 US service personnel as 'atomic veterans'. We don't know if their children are dying earlier than normal. The DVA has identified 195,000 US servicemen as being involved in the clean-up and occupation of Hiroshima and Nagasaki after they were bombed with atomic weapons at the end of World War II. A further 210,000 US service personnel participated in some 200 or so nuclear bomb tests conducted by the US military from 1945 until those tests were ended in 1963. 'Atomic veterans' have been exposed to various levels of radiation depending on the duties they performed. According to Drs C. Hansen and C. Schriner,[23] 'In 1995, the [DVA's] Institute of Medicine declined to recommend a study of the reproductive outcomes of Atomic Veterans.'[24]

◆

Before it broke up into individual nations in 1991, the Soviet Union had the largest number of 'radiation survivors' — men, women and children. This is due to the Chernobyl nuclear plant disaster in 1986.[25]

WHAT IS THE TRUTH ABOUT SCALPING?

It is generally assumed that scalping was unique to Native Americans, was practised only on white victims, involved slicing off all the hair and skin from the head of a victim, and all or nearly all scalping victims were dead or soon died. It is surprising to many that none of these statements is true. The custom of scalping has been found in numerous societies including parts of Europe. According to Dr Troy Case,[26] the origin of scalping in North America is unknown. It may have developed from the Native American custom of 'taking coup' from an enemy. Prestige came to a warrior who demonstrated bravery by touching the head of an enemy and escaping unharmed (taking coup). Prior to the arrival of Europeans, Native Americans scalped enemies infrequently. But the practice took place to some extent at least, because words exist in various Native American languages for the act of scalping, for the scalp itself and for the scalping victim that pre-date the arrival of Columbus. Another possibility is that scalping may have developed as a variant of the Native American warfare custom of cutting off various body parts as trophies.

Widespread scalping appeared only after the French and English settlers arrived in America. Two factors contributed to this change. First, the introduction to Native Americans of the steel knife made scalping physically easier to do, especially during battle. Second, high bounties were offered by the European colonists to various tribes for the scalps of members of other tribes. In 1703, Massachusetts colony paid bounties for scalps; as did New Hampshire colony in 1725. During the French and Indian War (1754–1760), the French paid for both the scalps of Native Americans and Europeans. This marks the beginning of the scalping of Europeans. Women and children were not exempt from being scalped. Before the bounties were introduced, scalping usually consisted of merely slicing off a small patch of skin from a victim who was either alive or dead. The scalp was only an inch or two in size and often it was taken from behind the ear. The scalping itself was rarely fatal. After the bounties were introduced, proof had to be given that the victim had been killed. Larger scalps were taken as proof ('No scalpy, no bounty!'). We can joke about it now, but it was all terribly gruesome then.[27, 28]

WHAT KILLED THE 'BUBBLE BOY'?
(Asked by Jenny Fredericks of Toronto, Ontario, Canada)

The film *Bubble Boy* (2001) is a romantic comedy in which a young man named Jimmy is born without an immune system and must therefore live his entire life within his bedroom inside a plastic bubble. The problem is he falls in love with Chloe. When Jimmy learns that Chloe is to be married in Niagara Falls, he builds a portable bubble suit and ventures into the big wide world in order to save his love. 'Bubble Boy Disease' is more correctly termed Severe Combined Immunodeficiency (SCID), which is actually a group of very rare congenital conditions.

Congenital means existing at birth and comes from the Latin *congenitus* meaning 'born together'.

SCID is very rare with only about 1 in every 500,000 to 1 in every million people suffering from it. People with SCID have severe abnormalities in both B- and T-cell immunity. This impaired immunity is in the endocrine system, but not so much in the nervous system. A person with SCID has an absent or almost absent supply of antibodies and white blood cells necessary to fight disease. This means that a person can die from their first serious infection. In order to protect themselves from just such an infection, they live in a sterile environment such as that provided by a plastic bubble. Rather like living in a fish bowl, but certainly better than dying. More technically, in the most common form of SCID, an X-linked chromosome problem results in a lack of the adenosine deaminase enzyme being produced. Without this enzyme, toxic body by-products such as ammonia accumulate in white blood cells and destroy them. SCID is not contagious. The symptoms usually show themselves in infants under the age of 3 months. The chief symptom of SCID is persistent infections, but a variety of other symptoms can also appear. SCID can be treated with intravenous immunoglobulin and bone marrow transplants. It is at least theoretically possible for a 'bubble boy' or 'bubble girl' to one day leave their bubble behind.

◆

The original 'bubble boys' were David Vetter (1971–1984) and Ted De Vita (1962–1980). Vetter suffered from SCID, lived in a plastic bubble and became almost a celebrity in his home town of Houston, Texas. De Vita did not have SCID. Instead, he suffered from aplastic anaemia, which nevertheless forced him to live for nearly 9 years in a sterile 'laminar air' hospital room at the US National Institutes of Health Clinical Centre in Bethesda, Maryland. Aspects of the lives of both boys were combined to inspire the film *The Boy in the Plastic Bubble* (1976), which helped make John Travolta a star in playing a character known as Tod Lubitch.[29]

CAN A CORPSE BURP?

(Asked by Gail Akami of Birmingham, UK)

A corpse does not burp if one means 'burp' in the usual sense of the belch of a living human. 'Corpse burping' confusion may come from one of two sources. First, in 1998, much attention was paid to the case of a supposed 'burping corpse'. According to the report, particularly popular in the US tabloid press at the time, allegedly the embalming of a Los Angeles man named Henry Galestor was postponed due to the mortician's observation that the dead man's stomach 'was convulsing every six or seven minutes'.[30] He supposedly described the convulsing as 'burping'. This observation was confirmed by a forensic scientist who supposedly wished to study the corpse. The story rather quickly achieved the status of urban legend. Second, those familiar with what happens when the human body dies are well aware of the fact that a decomposing body will undergo a liquefaction process as tissue breaks down by a combination of bacteria and enzyme action. These liquefied substances of a decaying body can be highly volatile. They can create gases. Under the right circumstances these gases can cause a small explosion. Knowing this, the funeral industry uses seals for the casket designed to give way somewhat before enough pressure builds inside to cause an explosion. In the funeral industry, this action of the seals giving way is sometimes euphemistically called 'burping'. No, a corpse does not 'burp' — nor will it say 'Excuse me' afterwards.[31]

IS FRIDAY THE 13TH BAD FOR YOUR HEALTH?

(Asked by Tanya Applegate of Ithaca, New York, USA)

Superstitions have existed since the beginning of humankind. Research backing such superstitions is another matter entirely. In 1988, Dr David Lester[32] wrote a brief article arguing that at least in the US there is a statistically significant greater number of suicides and homicides on Friday the 13th.[33] In 1993, Dr T.J. Scanlon and colleagues[34] presented evidence that in the UK the number of hospital admissions due to transport accidents increases by as much as 52 per cent on Friday the 13th. They write 'Friday [the] 13th is unlucky for some'. They even suggest that 'staying at home is recommended'.[35] In 2002, Dr Simo Nayha[36] examined Finland's death statistics from 1971 to 1997. According to Dr Nayha, this examination revealed that there was a statistically significant greater chance of dying in a traffic accident on Friday the 13th for women, but not for men. Dr Nayha went so far as to point out that 'an estimated 38% of traffic deaths involving women on this day were attributable to Friday the 13th itself'.[37] Just how this comes about was less clear.

Dr Nayha's findings were disputed by Dr Donald Smith[38] in the same journal.[39] Dr Nayha's findings were also challenged by Drs I. Radun and H. Summala.[40] These two researchers were very familiar with the statistics Dr Nayha had examined. After re-examining Dr Nayha's figures and investigating figures from Fridays the 13th from 1989 to 2002, Drs Radun and Summala concluded that there were 'no significant differences in any examined aspect of road injury accidents on Friday the 13th for either Finnish men or women'.[41] But that is not the 'finish' of this research. Dr V.V. Kumar and colleagues[42] investigated whether or not there is any truth to three beliefs often held by tonsillectomy specialists. The first is that more bleeding occurs in a tonsillectomy performed on a child with red hair. The second is that more bleeding occurs in a tonsillectomy performed during the day when there is a full moon at night. The third is that more bleeding occurs in a tonsillectomy performed on Friday the 13th. The three doctors conclude that no evidence exists to support any of the three beliefs. Despite all of this, the superstition will no doubt continue that Friday the 13th is bad for your health.[43, 44, 45]

WHAT IS MINAMATA DISEASE?

It is now 50 years since the most horrific mercury poisoning disaster the world has ever seen took place in Minamata, Japan. In May 1956, four patients from the city of Minamata on the west coast of the southern Japanese island of Kyushu were admitted to hospital with the same severe and baffling symptoms. They suffered from very high fever, convulsions, psychosis, loss of consciousness, coma and finally death. Soon afterwards, 13 other patients from fishing villages near Minamata suffered the same symptoms and also died. As time went on, more and more people became sick and many died. Doctors were puzzled by the strange symptoms and terribly alarmed. It was finally determined that the cause was mercury poisoning. Mercury was in the waste products dumped into Minamata Bay on a massive scale by a chemical plant. The mercury contaminated fish living in Minamata Bay, and the people who ate the fish were themselves contaminated and became ill. Local bird life and domestic animals also perished. In all, 900 people died and 2265 people were certified as having directly suffered from mercury poisoning — now known as Minamata disease.

Beyond this, victims who recovered were often socially ostracised as were members of their families. It was wrongly believed by many people in the community that the illness was contagious. The chemical plant was suspected of being the culprit in the environmental disaster almost from the beginning of the illness outbreak. Yet speaking out against the chemical plant was forbidden, because it was a major employer and enjoyed considerable economic and political clout all the way to the national government. Defenders of the chemical plant argued that they must be innocent since the plant had been in operation since 1907 without previous problems. It manufactured fertiliser.

A riot by local fishermen in 1959 finally moved the government to investigate the cause of the illnesses and deaths. Even so, it took officials 12 years from the first deaths to finally admit the cause of the contamination and order a halt to the mercury dumping into Minamata Bay. Still, the Minamata disaster story is not over. In 2006, Dr K. Eto[46] wrote that 'over the years, new facts have gradually surfaced, especially after 1995, with the resolution of the political problems surrounding Minamata disease'. For example, the mystery as to why the first 50 years of plant operation brought forth no disaster has been recently solved. It has been revealed that the plant modified its operations in August 1951 and started dumping large amounts of mercury directly into Minamata Bay only from that time. The health of survivors and their children is being monitored. A permanent museum and annual community ceremonies commemorate the worst mercury poisoning environmental disaster anywhere ever. Today, 50 years on, the lessons of Minamata remain.[47, 48]

◆

During his relatively long life, Sir Isaac Newton (1642–1727), perhaps the greatest scientist ever to have lived, suffered two serious bouts of uncharacteristically erratic behaviour. Some historians believe he suffered from a mild form of mercury poisoning. They point out that Newton was conducting experiments with mercury at the time of both occurrences.

◆

Mercury was used in the haberdashery industry into the 20th century. Hat makers were known to often suffer mental illnesses although the source of such illnesses was unknown. This is the basis of the name of the 'Mad Hatter' character in Lewis Carroll's *Alice in Wonderland* (1865).

CAN YOU DIE FROM TESTING A 9V BATTERY ON YOUR TONGUE?
(Asked by Liam Johnson of Frankfurt, Germany)

Here is Liam's question in full: 'First the simple question: Can you die from testing a 9v battery on your tongue? I have read newspaper reports (around 1990) of a woman dying after her boyfriend used a 9v battery as sexual stimulation. A brief look on the internet brings up a number of comments. 1) Apparently there was a US sailor being trained as an electrician who killed himself by sticking the probes of an Ohmmeter through his skin to measure his internal resistance. 2) There are also claims of 8 people a year dying in Australia (why just Australia?) from testing batteries. Now the explanation I have heard as to why this is, is that in certain freak cases, the battery can make an almost direct connection to the nervous system where the nerves are close to the surface and the skin is wet, thus ionizing the nerves so that they will not work correctly. Result is death. The detractors who attempt to answer this question invariably end up quoting figures for levels of current which kill then stating that a battery cannot generate these levels of current. This basically just dodges the issue since we have a fairly specific set of circumstances and it is not claimed that the current actually kills, rather the effect of a DC potentially connected directly to the nervous system. Besides, the figures quoted are guidelines for safety and not intended to be an absolute guarantee of safety. It is also my understanding, having spoken to a number of older electrical engineers, that there were different figures quoted for lethality for DC voltages and AC voltages, with the level of DC being significantly lower than for AC. Hope you can find some information on this issue! Thank you very much for your time!'

Dr Xheng Hu of the School of Electrical and Information Engineering at the University of Sydney confirms that a 9v battery does not have enough voltage to kill a person by testing it on the tongue. He adds, 'It cannot be entirely excluded however. If a person is very ill, for example, has heart problems, or has a heart pacemaker that could be disrupted, and so on, they could possibly die from testing the battery in this way. But normally it wouldn't happen.'[49, 50]

WHAT IS DIPHTHERIA?

(Asked by Andrew Wallace of Edinburgh, Scotland)

Most doctors in developed nations have never seen a diphtheria case. It's a wonderful thing that we've come so far. Sadly, diphtheria still regularly occurs in developing countries. Diphtheria is an extremely contagious and often life-threatening infection that usually attacks the membranes of the throat and nose. Although most often confined to the upper respiratory tract, in the most serious cases it attacks the nervous system and the heart. It is characterised by the formation of a tough membrane attached firmly to the underlying tissue that will bleed if forcibly removed. This is how diphtheria got its name, as *diphtheria* is the Greek word for leather. The infection begins in one tonsil and spreads to the other, then to the uvula, the soft palate, the pharyngeal wall, the larynx, the trachea and the bronchial tree. Bronchial obstruction occurs and death is due to suffocation.

Diphtheria is very easily spread by an infected person coughing, sneezing, or breathing on a person. Early symptoms are a sore throat, a fever and chills. Swallowing is difficult. But sometimes there are no symptoms for a considerable time. Yet during this time the infected person can spread the disease to others. Diphtheria is caused by *Corynebacterium diphtheriae*. If left untreated, this bacterium produces a powerful toxin that is carried in the bloodstream throughout the body. It can cause paralysis and heart failure. Diphtheria was once a greatly feared disease and far more widespread. In the US in the 1920s, it struck about 150,000 people a year. About 10 per cent died, with babies and children being particularly vulnerable.

◆

According to the US Centres for Disease Control and Prevention in Atlanta, Georgia, there were 918 diphtheria cases in 1960, 435 in 1970, 98 in 1980, 91 in 1990, and 88 in 2000.[51] Diphtheria is the 'D' in the DTP triple vaccine.

◆

An account exists of the last major diphtheria epidemic in Sydney, Australia, that took place in 1937. The disease advanced up the side of one street and down the other side killing nearly every child under the age of 5.[52]

♦

Leprosy is the world's oldest disease with documented cases dating back to 1350 BC. Leprosy is now 100 per cent curable.

♦

According to the World Health Organization (Geneva), measles ranks as the No. 1 disease worldwide that kills children and that also can be prevented through vaccination. In 2001, some 745,000 children died from measles. It is estimated that the cost of vaccinating all of these children would only have been just over half a million US dollars.

WHAT IS THE DIFFERENCE BETWEEN A VIRUS AND A BACTERIUM?
(Asked by Russell Stapleton of Melbourne, Victoria)

There are some subtle and not so subtle differences between a virus and a bacterium. A *virus* is an organism that contains no cells (acellular) whose genome consists of nucleic acid and that reproduces inside host cells. When they do this they use host metabolic machinery and ribosomes to form a pool of components which assemble into particles called virions. Virions serve to protect the genome and to transfer it to other cells. Viruses are distinct from other so-called virus-like agents such as viroids, plasmids and prions. Other characteristics of viruses are that they do not breathe, move or grow, however, they most definitely reproduce and can adapt to new hosts.

A *bacterium* is any of the one-celled (unicellular) prokaryotic micro-organisms of the class Schizomycetes. These vary in terms of their form and structure (morphology), oxygen and nutritional requirements, and capability to move (motility). They may be free-living, saprophytic (obtaining food by absorbing dissolved organic material), or capable of causing disease (pathogenic) in plants or animals. But bacteria are essential for human health too. For example, bacteria in the intestine are vital for digestion.[53]

IS THERE A MAXIMUM HEIGHT FROM WHICH A PERSON CAN SURVIVE A DIVE INTO WATER?
(Asked by Jack Crompton of Liverpool, UK)

The cliff divers of Acapulco amaze tourists with their bravery and skill. But from how much higher could they plunge without suffering serious injury or death? Surprisingly, it is difficult to be precise as to this height limit. While some unlucky ones die of falls in their bathtub, others have survived falls from incredible heights. For example, The Free Fall Research Page lists the case of World War II Russian airman, Lt I.M. Chisov. Chisov's Ilyushin IL–4 bomber was shot down by German fighters in January 1942. Chisov fell 22,000 feet (6705 m), hit the edge of a snow-covered ravine and rolled to the bottom. Despite being badly injured, Chisov survived. Although this is not cliff diving into water, it shows what is possible.

Intricately involved in any such calculation of maximum survival height is terminal velocity. Terminal velocity is the maximum speed of free fall of a human in air. Once terminal velocity is reached, no matter how much higher one falls from, they will not increase their speed in falling. Although there is some dispute about this figure, the terminal velocity of a human is estimated to be about 325 km per hour. The speed of a diver from a 30 metre cliff is estimated to be only about 90 km per hour. This is only one-third or so of the terminal velocity. Another factor too is plunging position. If the diver is plunging head-first, their speed will be somewhat faster than if they were falling spread-eagled due to less drag in the head-first position. According to Linn Emrich, author of *The Complete Book of Sky Sports* (1970), a 77 kg person would reach terminal velocity after about

14 seconds. They would fall nearly 10,000 feet (3048 m) in 1 minute. Cliff divers are not in the air for anywhere near 14 seconds. This is why they can dive and survive.

Some living creatures have a terminal velocity that is not fatal. For example, ants can survive falls from heights that would be easily fatal for humans. But cliff-diving ants would not be nearly as popular with tourists in Acapulco.[54, 55, 56]

WHAT HAPPENS WHEN YOU ARE EXECUTED BY ELECTROCUTION?
(Asked by Ron Talbot of Tyler, Texas, USA)

In the late 19th century, it was widely believed that a more modern method of execution was needed to replace the three most commonly used execution methods at that time — hanging, firing squad and, in France, beheading. The first practical electric chair was invented by Harold P. Brown who worked for Thomas Edison. The first person to die in the electric chair was executed in 1890. A parallel occurrence at around the same time was the scientific discovery of the precise effects on the body of high voltages of electricity. For example, according to Dr T. Bernstein,[57] two doctors by the name of Prevost and Battelli demonstrated in 1899 that death from electrocution was caused, not by damaging the brain, but by high voltages of electricity causing very rapid irregular contractions of the heart (ventricular fibrillation) eventuating in the heart stopping. As for the execution itself, the prisoner must first be prepared for execution by shaving the head and the calf of one leg. This permits better contact between the skin and the electrodes which must be attached to the body. The prisoner is strapped into the electric chair at the wrists, waist and ankles. An electrode is attached to the head and another to the leg. At least two jolts of an electrical current are applied for several minutes. An initial voltage of about 2000 volts stops the heart and induces unconsciousness.

The voltage is then lowered somewhat. In one US state, the protocol calls for a jolt of 2450 volts that lasts for 15 seconds. After a 15 minute wait, the prisoner is then examined by a coroner. After 20 seconds, the cycle is repeated three more times. The body may heat up to approximately 100°C. which causes severe damage to internal organs. Often the eyeballs melt. Taping the eyes closed is usually part of the preparation for execution by electrocution. The effects of the electricity often cause the body to twitch and gyrate uncontrollably and bodily functions may 'let go'. Prisoners are sometimes offered diapers.

Although death is supposedly instantaneous, some prisoners have been known to shriek and even shout while being executed in this way. There have been reports of a prisoner's head bursting into flames. There have been reports too of a prisoner being removed from an electric chair that has malfunctioned part way through the electrocution and then being placed back in the chair once it was fixed in order for the job to be finished. Some skin that is burned off the prisoner must then be scraped off the seat and straps of the electric chair before it may be used again.[57]

♦

In 1991, a recommendation was made by two Polish doctors that the thighs also be strapped to the chair. Drs L. Zynda and K. Skiba of Warsaw reported on the case of a 58-year-old executed male whose legs were broken by the intense twitching due to the force of the deadly electric current passing through his body.[58]

♦

In 1946, an electric chair malfunctioned and failed to execute the prisoner who reportedly shrieked, 'Stop it! Let me breathe!' as he was being executed. Having survived, lawyers for the prisoner argued that, although he did not die, he had been executed as defined by the law. In 1947, in the case of Francis vs Resweber, the US Supreme Court ruled against the prisoner. He was returned to the electric chair and successfully executed later that year.[59, 60]

IS IT TRUE THAT MORE US CITIZENS DIE FROM PEANUTS THAN FROM TERRORISM?

(Asked by Lance Jones of Nashville, Tennessee, USA)

As strange as it may seem, more US citizens die each year from eating peanuts than from acts of terrorism. According to the US Department of State's Office of the Coordinator for Counterterrorism, in 2005, 56 US citizens worldwide were killed as a result of terrorism.[61] This compares to about 100 people in the US who die each year from eating peanuts.

A serious food allergy is called food anaphylaxis. Food anaphylaxis is now the leading known cause of all anaphylactic reactions treated in emergency departments throughout the US. It is estimated that there are 30,000 cases each year resulting in some 150 to 200 deaths. Anaphylaxis is an inflammatory reaction that leads to damaging outcomes to the organism. It is the opposite of prophylaxis which *prevents* damaging outcomes to the organism. An anaphylactic reaction typically consists of shortness of breath, rash, wheezing and unusually low blood pressure (hypotension). In comparatively rare cases, an anaphylactic reaction results in death. According to Dr H.A. Sampson,[62] 'peanuts, tree nuts, fish and shellfish account for most severe food anaphylactic reactions'. Inconsistent coding of deaths following World Health Organization (WHO) categories makes statistics somewhat unreliable as to precisely how many peanut deaths occur each year in the US and to US nationals living overseas.[63]

◆

According to the WHO in Geneva, in 2002 there were four people killed worldwide in traffic accidents for every one person killed in war.[64]

◆

Thirteen people die each year worldwide from vending machines falling on them.

◆

The chances of being killed falling out of bed are 1 in 2 million.

◆

About 60,000 deaths worldwide occur each year from over exposure to the sun.

WHAT HAPPENS TO PEOPLE WHEN THEY ARE STRUCK BY LIGHTNING?
(Asked by Felix Verity of Manila, The Philippines)

It is estimated that each year just in the US, at least 600 people are injured by lightning and about 2500 are injured by electric shock. Of course, when people are struck by lightning they can die. But what is far more interesting is the type of injuries that can be sustained and the temporary and permanent changes to the body that can occur from being struck by lightning. The Lightning Strike and Electric Shock Survivors International organisation is a support group located in Jacksonville, North Carolina. It boasts members in 13 nations besides the US and has an annual convention where survivors share their 'shocking' stories. The publications of the organisation include two *Life After Shock* volumes and a third volume in the series that is 'coming soon'.

The accounts of survivors are varied. There are typical cases where victims now must walk with canes, have artificial limbs, are now confined to wheelchairs, or suffer nervous tics, stuttering, memory loss, depression, blurred vision, loss of hearing or impotence. After all, the human body has been exposed to a tremendously high level of electrical voltage. It is said that 'although the body's hardware may not be damaged, the body's software certainly is'. But then there are the extraordinary tales. For example, one man was struck by lightning and has since become impervious to frigid weather. He delights in romping through snow in his swimming costume and taking a bath in ice water. Another man could speak 11 languages before his encounter with 3900 volts as a result of accidentally touching a broken outlet. He lost his foreign language ability altogether, stammers through a simple conversation in English and has no sense of direction. Still another man was a Kung Fu grandmaster who ran 16 km (10 miles) and did 300 push-ups per day. After being struck by lightning he can only do about five push-ups without getting dizzy.

◆

Most deaths that occur as a result of being struck by lightning are due to cardiac arrest.

◆

The state of Florida leads the US in the number of deaths and injuries due to lightning strikes.

◆

It is estimated that 10 per cent of lightning strike cases in the US go unreported.

◆

About half of people struck by lightning lose their jobs for one reason or another.

◆

About one half of married people struck by lightning get divorced for reasons at least in part having something to do with the lightning strike.[65]

◆

Does lightning strike a human twice? The only human to be struck seven times by lightning was Roy Sullivan, a park ranger from the US state of Virginia. The strikes occurred between 1942 and 1977. Ironically, having survived these strikes, 'the human lightning rod' committed suicide in 1983, reputedly over a failed romance.

Afterword

Keep those questions coming in.

If you have an unusual curly question (or two) about the body that has not been answered in earlier books, send it in and we will try to answer it and possibly publish it in the next book. We will even send you a free copy of the book. The books we have given away so far have gone to every continent, including Antarctica. If we have answered the question already, we still give you a personal response.

Send your question to:

Dr Stephen Juan

c/o HarperCollins Publishers Australia

25 Ryde Road

Pymble NSW 2073 AUSTRALIA

OR

Email to: drstephenjuan@exemail.com.au

References

Note: sources from the *National Post* in Toronto are from Dr Stephen Juan's column 'Body and Health'; sources from *The Register* in London and the *Epoch Times* in Sydney are from his columns 'The Odd Body'.

Chapter 1: Beginnings

1 C. Ray, 'How is paternity determined and with what degree of accuracy?', *New York Times*, 9 December 1986, p. B4.
2 C. Sutton, *How Did They Do That?*, Quill, New York, 1985, pp. 91–94.
3 E. Main, D. Moore, B. Farrell, L. Schimmel, R. Altman, C. Abrahams, M. Bliss, L. Polivv and J. Sterling, 'Is there a useful cesarean birth measure? Assessment of the nulliparous term singleton vertex cesarean birth rate as a tool for obstetric quality improvement', *American Journal of Obstetrics and Gynecology*, 2006, vol. 194, no. 6, pp. 1644–1651.
4 W. Burket, *Creation of the Sacred: Tracks of Biology in Early Religions*, Harvard University Press, Cambridge, Massachusetts, 1996.
5 A. Mohl, 'Growing up male: Is violence, crime and war endemic to the male gender?', *Journal of Psychohistory*, 2006, vol. 33, no. 3, pp. 270–289.
6 S. Juan, 'Defining race', *National Post*, 10 April 2006, p. 1.
7 D. Jones, 'The Neanderthal code?', *New Scientist*, 11 November 2006, pp. 44–47.
8 S. Guynup, 'Resurrecting extinct animals?', *Popular Science*, February 2006, pp. 54–55.
9 S. Juan, 'Bringing back the extinct', *National Post*, 1 May 2006, p. 1.
10 S. Juan, 'What is the difference between a chromosome and a gene?', *The Register*, 19 May 2006.
11 S. Juan, 'What are chromosome abnormalities and how often do they occur?', *The Register*, 19 May 2006.
12 Dr Michael De Bellis is from the University of Pittsburgh Medical Centre.
13 L. Thomas and M. De Bellis, 'Pituitary volumes in pediatric maltreatment-related posttraumatic stress disorder', *Biological Psychiatry*, 2004, vol. 55, no. 7, pp. 752–758.
14 Dr Martin Teicher is from the Developmental Biopsychiatry Research Program at McLean Hospital, Harvard University.
15 M. Teicher, N. Dumont, Y. Ito, C. Vaituzis, J. Giedd and S. Andersen, 'Childhood neglect is associated with reduced corpus callosum area', *Biological Psychiatry*, 2004, vol. 56, no. 2, pp. 80–85.

16   Dr Margot Sunderland is the Director of Education and Training at the Centre for Child Mental Health in London.

17   S. Juan, 'Can leaving a baby to "cry it out" cause brain damage?', *The Register*, 14 July 2006.

18   S. Juan, 'Can leaving my baby to "cry it out" cause brain damage?', *National Post*, 30 October 2006, p. 1.

19   Anni Gethin is a health social scientist in Sydney and Beth Macgregor is a psychologist in Sydney.

20   A. Gethin and B. Macgregor, *Helping Your Baby to Sleep*, Finch Publishing, Sydney, 2007, p. 51.

21   Dr Jeffry Simpson is from the Institute of Child Development at the University of Minnesota in Minneapolis.

22   J. Simpson, W. Collins, S. Tran and K. Haydon, 'Attachment and the experience and expression of emotions in romantic relationships: A developmental perspective', *Journal of Personality and Social Psychology*, 2007, vol. 92, no. 2, pp. 355–367.

23   C. Ray, 'What is amniotic fluid?', *New York Times*, 12 January 1999, p. D3.

24   S. Juan, 'What is amniotic fluid?', *The Register*, 18 August 2006.

25   S. Juan, 'Great moments in human research', *The Register*, 27 January 2007.

26   S. Juan, 'Great moments in human research', *The Register*, 3 February 2007.

27   Drs Anthony DeCasper and Melanie Spence are from the University of North Carolina at Greensboro.

28   A. DeCasper and M. Spence, 'Prenatal maternal speech influences newborns' perception of speech sounds', *Infant Behaviour and Development*, 1986, vol. 9, no. 2, pp. 133–250.

29   D. Chamberlain, *The Mind of Your Newborn Baby*, North Atlantic Books, Berkeley, 1998, pp. 37–38.

30   S. Juan, 'Why can't I remember my own birth?', *The Register*, 8 September 2006.

31   Drs K.Y. Loh and N. Sivalingam are from the International Medical University in Kuala Lumpur, Malaysia.

32   K. Loh and N. Sivalingam, 'Understanding hyperemesis gravidarum', *Medical Journal of Malaysia*, 2005, vol. 60, no. 3, pp. 394–399.

33   Drs J.D. Quinla and D.A. Hill are from the naval hospital in Jacksonville, Florida.

34   J. Quinla and D. Hill, 'Nausea and vomiting of pregnancy', *American Family Physician*, 2003, vol. 68, no. 1, pp. 121–128.

35   Drs Gillian Pepper and S. Craig Roberts are from the School of Biological Sciences at the University of Liverpool.

36   G. Pepper and S. Roberts, 'Rates of nausea and vomiting in pregnancy and dietary characteristics across populations', *Proceedings of the Royal Society, Biological Sciences*, 2006, vol. 273 (1601), pp. 2675–2679.

37   Dr C. Paquin is a biologist at the University of Laval in Quebec, Canada, and Dr J. Adams is a biologist at the University of Michigan in Ann Arbor.

38   C. Paquin and J. Adams, 'Frequency of fixation of adaptive mutations is higher in evolving diploid than haploid yeast populations', *Nature*, 1983, vol. 302 (5908), pp. 495–500.

39 C. Paquin and J. Adams, 'Relative fitness can decrease in evolving asexual
    populations of *S. cerevisiae*', *Nature*, 1983, vol. 306 (5941), pp. 368–370.
40 S. Juan, 'Why does natural selection take so long to get results?', *The Register*,
    15 September 2006.
41 Dr Ralph Catalano is a professor of public health at the University of California
    at Berkeley.
42 R. Catalano, 'Sex ratios in the two Germanies: A test of the economic stress
    hypothesis', *Human Reproduction*, 2003, vol. 18, no. 9, pp. 1972–1975.
43 R. Catalano, T. Bruckner, A. Marks and B. Eskenazi, 'Exogenous shocks to the
    human sex ratio: The case of September 11, 2001 in New York City', *Human
    Reproduction*, 2006, vol. 21, no. 12, pp. 3127–3131, Epub 26 August 2006.
44 R. Catalano and T. Bruckner, 'Male lifespan and the secondary sex ratio',
    American Journal of Human Reproduction, 2006, vol. 18, no. 6, pp. 783–790, Epub
    12 October 2006.
45 'Fewer boys are born during hard times', *New Scientist*, 30 August 2006, p. 20.
46 S. Juan, 'Is it true that fewer boy babies are born in hard times?', *The Register*,
    3 November 2006.
47 S. Juan, 'What is Cro-Magnon man?', *The Register*, 10 November 2006.
48 Dr Jimmy Or is from the Takanishi Laboratory Humanoid Robotics Institute of
    Waseda University in Tokyo.
49 J. Or, 'A control system for a flexible spine belly-dancing humanoid', *Artificial Life*,
    2006, vol. 12, pp. 63–87.
50 S. Juan, 'Meet the belly dancing robot', *National Post*, 8 January 2007, pp. 1–2.
51 Drs M. Hirose and K. Ogawa are from Honda Research and Development
    Company Ltd of the Wako Research Centre in Saitama, Japan.
52 M. Hirose and K. Ogawa, 'Honda humanoid robots development', *Philosophical
    Transactions of the Royal Society A: Mathematical, Physical and Engineering Sciences*,
    2007, vol. 365 (1850), pp. 11–19.
53 Drs A. Arita, K. Hiraki, T. Kanda and H. Ishiguro are from the Department of
    General Systems Studies at the University of Tokyo.
54 A. Arita et al., 'Can we talk to robots? Ten-month-old infants expected interactive
    humanoid robots to be talked to by persons', *Cognition*, 2005, vol. 95, vol. 3,
    pp. 849–857.
55 Dr Hiroshi Ishiguro is from the ATR Intelligent Robotics and Communication
    Laboratories, near Kyoto, Japan.
56 B. Schaub, 'My android twin', *New Scientist*, 14 October 2006, pp. 42–46.
57 C. Biever, 'A good robot has personality but not looks', *New Scientist*, 22 July 2006,
    p. 32.
58 Drs K. Nishiwaki, J. Kuffner, S. Kagami, M. Inaba and H. Inoue are from the
    Digital Human Research Centre of the National Institute of Advanced Industrial
    Science and Technology of Tokyo.
59 K. Nishiwaki et al., 'The experimental humanoid robot H7: A research
    platform for autonomous behaviour', *Philosophical Transactions of the Royal
    Society A: Mathematical, Physical and Engineering Sciences*, 2007, vol. 365 (1850),
    pp. 79–107.
60 Dr Alain Cardon is from the Laboratory of Information of Paris.

61 A. Cardon, 'Artificial consciousness, artificial emotions, and autonomous robots', *Cognitive Processes*, 2006, vol. 7, no. 4, pp. 245–267.

62 R. Kurzweil, 'Robots R Us', *Popular Science*, September 2006, pp. 52–71.

63 S. Juan, 'Will robots ever become just like humans?', *The Register*, 23 December 2006.

64 S. Juan, 'How old is my body if the cells keep renewing themselves?', *The Register*, 17 February 2007.

65 Dr Michael Onken is from the Department of Ophthalmology and Visual Sciences at Washington University in St Louis.

66 Personal communication, 2 February 2006.

67 Dr Barbara Sakakian is from the Department of Psychiatry at the School of Clinical Medicine of Cambridge University.

68 Personal communication, 7 February 2006.

69 Dr Aubrey de Grey is from the Interdisciplinary Research Centre on Aging at Cambridge University.

70 G. Lawton, 'The incredibles', *New Scientist*, 15 May 2006, pp. 32–38.

Chapter 2: The Head

1 S. Juan, 'Is the human skull made up of one bone or two?', *The Register*, 20 January 2007.

2 Drs A. Czaplinski, A. Steck and P. Fuhr are from the Neurology Clinic of the University of Bazylei in Szwajacaria, Poland.

3 A. Czaplinski, A. Steck and P. Fuhr, 'Tic syndrome', *Neurologia neurochirurgia polska*, 2002, vol. 36, no. 3, pp. 493–504.

4 Dr Alumit Ishai is from the Institute of Neuroradiology at the University of Zurich in Switzerland.

5 A. Ishai, 'Sex, beauty and the orbitofrontal cortex', *International Journal of Psychophysiology*, 2007, vol. 63, no. 2, pp. 181–185.

6 Dr David Perrett is a cognitive psychologist at the University of St Andrews in Scotland.

7 D. Perrett, K. Lee, I. Penton-Voak, D. Rowland, S. Yosikawa, D. Burt, S. Henzi, D. Castles and S. Akamatsu, 'Effects of sexual dimorphism on facial expression', *Nature*, 1998, vol. 394 (6696), pp. 884–887.

8 S. Juan, 'Why is it that we find some faces so attractive and not others?', *National Post*, 19 April 2007, pp. 1–2.

9 Dr Arthur Aron is from the Department of Psychology at the State University of New York at Stony Brook.

10 A. Aron, H. Fisher, D. Mashek, G. Strong, H. Li and L. Brown, 'Reward, motivation and emotion systems associated with early-stage intense romantic love', *Neurophysiology*, 2005, vol. 94, no. 1, pp. 327–337.

11 H. Fisher, A. Aron and L. Brown, 'Romantic love: A mammalian brain system for mate choice', *Philosophical Transactions of the Royal Society B: Biological Sciences*, 2006, vol. 361 (1476), pp. 2173–2186.

12 S. Juan, 'Great moments in human research', *The Register*, 27 January 2007.

13 S. Juan, 'Great moments in human research', *The Register*, 3 February 2007.

14 Drs J.R. Meloy and H. Fisher are from the Department of Psychiatry of the University of California at San Diego.

15   J. Meloy and H. Fisher, 'Some thoughts on the neurobiology of stalking', *Journal of Forensic Sciences*, 1995, vol. 50, no. 6, pp. 1472–1480.

16   Dr N.C. Heglund is from the Catholic University of Louvain in Belgium.

17   N. Heglund, P. Willems, M. Penta and G. Cavagna, 'Energy-saving gait mechanics with head-supported loads', *Nature*, 1995, vol. 375 (6526), pp. 52–54.

18   J. Varasdi, *Myth Information: More than 590 Popular Misconceptions, Fallacies and Misbeliefs Explained*, Ballantine, New York, 1996, p. 122.

19   J. Rains and F. Taylor, *Chronic Daily Headache: An Overview*, American Council for Headache Education, Mount Royal, New Jersey, 2006.

20   Mayo Clinic, *Thunderclap Headaches*, Mayo Clinic Foundation for Medical Education, Rochester, Minnesota, 24 May 2005.

21   S. Juan, 'Aching heads', *National Post*, 10 April 2006, p. 1.

22   Mayo Clinic, *Ice Cream Headaches*, Mayo Clinic Foundation for Medical Education, Rochester, Minnesota, 24 May 2006.

23   S. Juan, 'What is a brain freeze?', *The Register*, 16 June 2006.

24   Johns Hopkins Medicine, *Health Information Library: MSG Headaches*, Johns Hopkins Health Information Service, Baltimore, Maryland, 24 May 2006.

25   S. Juan, 'What is a Chinese restaurant headache?', *The Register*, 16 June 2006.

26   S. Juan, 'Why doesn't a hangover occur the night before?', *The Register*, 28 July 2006.

27   S. Juan, 'Your head will hurt tomorrow', *National Post*, 13 November 2006, p. 1.

28   H. Moore, 'Avoiding post-lumbar puncture headaches', *Pulmonary Reviews*, 2000, vol. 5, no. 12, pp. 1–7.

29   Dr R. Gaiser is from the Department of Anesthesiology at the University of Pennsylvania.

30   R. Gaiser, 'Postdural puncture headache', *Current Opinion in Anaesthesiology*, 2006, vol. 19, no. 3, pp. 249–253.

31   Dr C.L. Wu is from the Department of Anesthesiology at Johns Hopkins University.

32   C. Wu, A. Rowlingson, S. Cohen, R. Michaels, G. Courpas, E. Joe and S. Liu, 'Gender and post-dural puncture headache', *Anesthesiology*, 2006, vol. 105, no. 3, pp. 613–618.

33   S. Juan, 'What is a post-lumbar headache?', *The Register*, 27 October 2006.

34   Dr Andrew Lloyd is an infectious disease physician in Sydney, Australia.

35   Personal communication, 29 August 2006.

36   Dr J.D. Grabenstein is from the US Office of the Surgeon General.

37   J. Grabenstein, P. Pitman, J. Greenwood and R. Engler, 'Immunization to protect the US Armed Forces: Heritage, current practice, and prospects', *Epidemiologic Reviews*, 2006, vol. 28, no. 1, pp. 3–26.

38   S. Juan, 'Can you get tetanus from a rusty nail?', *The Register*, 30 June 2006.

39   Dr A. Prysyazhnyuk is from the Research Centre for Radiation Medicine of AMS of the Ukraine in Kiev.

40   A. Prysyazhnyuk, V. Gristchenko, Z. Fedorenko, L. Gulak, M. Fuzik, K. Slipenyuk and M. Tirmarche, 'Twenty years after the Chernobyl accident: Solid cancer incidence in various groups of the Ukrainian population', *Radiation and Environmental Biophysics*, 2007, vol. 46, no. 1, pp. 43–51.

41   Drs R.A. Schwartz, C.A. Janusz and C.K. Janniger are from the University of Medicine and Dentistry at the New Jersey Medical School in Newark.

42   R.A. Schwartz et al., 'Seborrheic dermatitis: An overview', *American Family Physician*, 2006, vol. 74, no. 1, pp. 125–130.

Chapter 3: The Eyes

1   Dr Brent Archinal is from the Astrogeology Team of the US Geological Survey in Flagstaff, Arizona.

2   B. Archinal, 'How far can you see?', *Astronomy*, May 1997, p. 20.

3   J. Apt, 'Orbit, the astronauts' view of home', *National Geographic*, November 1996, pp. 8–27.

4   D. Fisk and R. Brown, 'Chinese puzzle', *New Scientist*, 15 July 1995, p. 65.

5   S. Juan, 'How far can the naked eye see?', *The Register*, 1 December 2006.

6   S. Juan, 'Can the Great Wall be seen from the moon?', *Epoch Times*, 29 November 2006.

7   S. Juan, 'Can you really see the Great Wall of China from the Moon?', *The Register*, 1 December 2006.

8   Dr Samuel Salamon is from the Cataract Eye Centre of Cleveland, Ohio.

9   S. Juan, 'Why do babies blink less often than adults?', *The Register*, 30 June 2006.

10   D.M. Stein, G. Wollstein, H. Ishikawa, E. Hertzmark, R. Noecker and J. Schuman are from the Department of Ophthalmology at the School of Medicine at the University of Pittsburgh.

11   D. Stein et al., 'Effect of corneal drying on optical coherence tomography', *Ophthalmology*, 2006, vol. 113, no. 6, pp. 985–991.

12   Dr Stephen Miller is the Director of the Clinical Care Centre of the American Optometric Association in St Louis.

13   Personal communication, 23 August 2006.

14   Drs N.S. Logan, L. Davies, E. Mallen and B. Gilmartin are from the Human Myopia Research Centre at Aston University in Birmingham, UK.

15   N. Logan et al., 'Ametropia and ocular biometry in a UK university student population', *Optometry and Vision Science*, 2005, vol. 82, no. 4, pp. 261–266.

16   S. Juan, 'Why are so many humans near sighted?', *The Register*, 22 September 2006.

17   S. Juan, 'How does a cross-eyed person's view differ from others?', *The Register*, 6 October 2006.

18   Dr Michael Lawless of the Department of Ophthalmology at the Royal North Shore Hospital in Sydney, Australia.

19   Personal communication, 26 August 2006.

20   Drs M.A. Bullimore, K. Reuter, L. Jones, G. Mitchell, J. Zoz and M. Rah are from the Ohio State University College of Optometry in Columbus.

21   M. Bullimore et al., 'The study of progression of adult nearsightedness (SPAN): Design and baseline characteristics', *Optometry and Vision Science*, 2006, vol. 83, no. 8, pp. 594–604.

22   Dr Armand Tanguay Jr is a professor of electrical engineering at the University of Southern California.

23   M. Stroh, 'We see the future better than 20/20', *Popular Science*, June 2005, p. 59.

24   S. Juan, 'Is an artificial eye close to reality?', *The Register*, 24 November 2006.
25   'Blind person sees colour with touch only', *Dominican Today* (Santo Domingo), 17 November 2005, p. 1.
26   S. Juan, 'Can the blind feel colours', *Epoch Times*, 15 November 2006, p. 11.
27   Dr Daniel J. Simons is from the Department of Psychology at Harvard University.
28   D. Simons, 'Attentional capture and inattentional blindness', *Trends in Cognitive Sciences*, 2000, vol. 4, no. 4, pp. 147–155.
29   Drs S.B. Most, B. School, E. Clifford and D. Simons are from the Department of Psychology at Harvard University.
30   S. Most et al., 'What you see is what you set: Sustained inattentional blindness and the capture of awareness', *Psychological Review*, 2005, vol. 112, pp. 217–242.
31   Drs Mika Koivisto and Antti Revonsuo are from the Centre for Cognitive Neuroscience at the University of Turku in Finland.
32   M. Koivisto and A. Revonsuo, 'The role of unattended distracters in sustained inattentional blindness', *Psychological Research*, 2008, vol. 72, no. 1, pp. 39–48, Epub 5 July 2006.
33   S. Juan, 'When is seeing not seeing?', *The Register*, 19 January 2006.
34   Dr Kenton McWilliams is from the School of Optometry at the University of Missouri in St Louis.
35   Personal communication, 18 May 2006.
36   R. Williams and W. Madil, 'Goggle eyed', *New Scientist*, 17 June 2000, p. 65.
37   Dr Anna Gislen is from Lund University in Sweden.
38   A. Gislen and L. Gislen, 'On the optical theory of underwater vision in humans', *Journal of the Optical Society of America A: Optics, Image Science, and Vision*, 2004, vol. 21, no. 11, pp. 2061–2064.
39   A. Gislen, E. Warrant, M. Dacke and R. Kroeger, 'Visual training improves underwater vision in children', *Vision*, 2006, vol. 46, no. 20, pp. 3443–3450.
40   J. Travis, 'The eyes have it', *Science News*, 17 May 2003, p. 308.
41   Drs Eugene Aserinsky and Nathaniel Kleitman were at the University of Chicago at the time of their research.
42   E. Aserinsky and N. Kleitman, 'Regularly occurring periods of eye motility, and concomitant phenomena, during sleep', *Science*, 1953, vol. 118 (3062), pp. 273–274.
43   Dr David Maurice is from Columbia University.
44   D. Maurice, 'The Von Sallmann lecture 1996: An ophthalmological explanation of REM sleep', *Experimental Eye Research*, 1996, vol. 66, no. 2, pp. 139–145.
45   Drs F. Hoffmann and G. Curio are from the Free University of Berlin.
46   F. Hoffmann and G. Curio, 'REM sleep and recurrent corneal erosion — hypothesis', *Klinische Monatsblatter fur Augenheilkunde*, 2003, vol. 220, nos. 1–2, pp. 51–53.
47   'Rolling eyes gather more oxygen', *New Scientist*, 28 February 1998, p. 23.
48   A. Mijolla, *International Dictionary of Psychoanalysis*, eNotes, Seattle, 26 May 2006.
49   G. Cook, 'How do we take advantage of inflection points?', *Cook & Company Commentary*, Winter 2003, p. 2.
50   S. Juan, 'What's this "scotomisation" in *The Da Vinci Code*?', *The Register*, 9 June 2006.

51   S. Juan, 'Why seeing is not always believing', *National Post*, 28 September 2006, pp. 1–2.

52   S. Juan, 'Great moments in human research', *The Register*, 27 January 2007.

53   S. Juan, 'Great moments in human research', *The Register*, 3 February 2007.

Chapter 4: The Nose

1   University of California, San Diego Medical Centre, *Types of Nasal Dysfunction*, University of California, San Diego, 12 August 2006.

2   S. Juan, 'Are there people with no sense of smell?', *The Register*, 16 September 2006.

3   S. Juan, 'Is there an evolutionary advantage in snoring?', *The Register*, 14 July 2006.

4   S. Juan, 'What evolutionary advantage is there in making a sound while snoring?', *National Post*, 30 October 2006, pp. 1–2.

5   Drs J.A. Gottfried and R.J. Dolan are from the Functional Imaging Laboratory of the Wellcome Department of Imaging Neuroscience in London.

6   J. Gottfried and R. Dolan, 'The nose smells what the eye sees: Crossmodal visual facilitation of human olfactory perception', *Neuron*, 2003, vol. 39, no. 2, pp. 375–386.

7   Drs R.A. Osterbauer, P. Matthews, M. Jenkinson, C. Beckmann, P. Hansen and G. Calvert, are from the Oxford Centre for Functional Magnetic Resonance Imaging of the Brain at Oxford University.

8   R. Osterbauer et al., 'Colour of scents: Chromatic stimuli modulate odour responses in the human brain', *Journal of Neurophysiology*, 2005, vol. 93, no. 6, pp. 3434–3441.

9   Drs S. Lombion-Pouthier, P. Vandel, S. Nezelof, E. Haffen and J. Millot are from the Laboratoire de Neurosciences at the Université de Franche-Comté in Cedex, France.

10   S. Lombion-Pouthier et al., 'Odor perception in patients with mood disorders', *Journal of Affective Disorders*, 2006, vol. 90, nos. 2–3, pp. 187–191.

11   Drs K. Sugiyama, Y. Hasegawa, N. Sugiyama, M. Suzuki, N. Watanabe and S. Murakami are from the Nagoya City University Medical School.

12   K. Sugiyama et al., 'Smoking-induced olfactory dysfunction in chronic sinusitis and assessment of brief University of Pennsylvania Smell Identification Test and T&T methods', *American Journal of Rhinology*, 2006, vol. 20, no. 5, pp. 439–444.

13   S. Juan, 'Who knows what there is to know about the nose?', *The Register*, 17 November 2006.

14   Dr Betty Repacholi is from the Department of Psychology at the University of Washington in Seattle.

15   T. Case, B. Repacholi and R. Stevenson, 'My baby doesn't smell as bad as yours: The plasticity of disgust', *Evolution and Human Behaviour*, 2006, vol. 27, no. 5, pp. 357–365.

16   Drs V. Curtis, R. Aunger and T. Rabie are from the London School of Hygiene and Tropical Medicine.

17   V. Curtis et al., 'Evidence that disgust evolved to protect from risk of disease', *Proceedings/Biological Sciences. The Royal Society*, 2004, vol. 271, suppl. 4, pp. S131–133.

18   R. Rose and J. Resnick, 'Waste disposal?', *New Scientist*, 19 August 2006, p. 57.

19   S. Juan, 'Who knows what there is to know about the nose?', *The Register*, 17 November 2006.

20   V. Iannelli, *Do Babies Have Sinuses?*, Your Guide to Pediatrics, About Inc., New York, 4 July 2006.

21   S. Juan, 'Why do babies always seem to have a runny nose?', *The Register*, 28 July 2006.

22   S. Juan, 'Why do babies often have a runny nose?', *National Post*, 13 November 2006, p. 1.

23   S. Juan, 'Great moments in human research', *The Register*, 27 January 2007.

24   S. Juan, 'Great moments in human research', *The Register*, 3 February 2007.

25   D. Feldman, *What are Hyenas Laughing at, Anyway?*, HarperCollins, New York, 1996, p. 61.

26   Dr Noam Sobel is now a professor of psychology at the Helen Wills Neuroscience Institute of the University of California at Berkeley.

27   M. Barraud, 'Two sides of it', *New Scientist*, 6 November 1999, p. 6.

28   L. Watson, *Jacobson's Organ and the Remarkable Nature of Smell*, W.W. Norton, New York, 2000.

29   L. Lowndes, *How to Make Anyone Fall in Love With You*, McGraw-Hill, New York, 1997, p. 293.

30   S. Juan, 'Vomeronasal organ: Dead or alive?', *The Register*, 9 May 2006.

31   Drs D.M. Bautista, P. Movahed, A. Hinman, H. Axelsson, O. Sterner, E. Hogestatt, D. Julius, S. Jordt and P. Zygmunt are from the Department of Cellular and Molecular Pharmacology at the University of California in San Francisco.

32   D. Bautista et al., 'Pungent products from garlic activate the sensory ion channel', *Proceedings of the National Academy of Science*, 2005, vol. 102, no. 34, pp. 12248–12252.

33   T. Mendham, *Garlic Breath*, Garlic Central, Edinburgh, 9 November 2006.

34   S. Juan, 'Who knows what there is to know about the nose?', *The Register*, 17 November 2006.

35   Dr Hans Wohlmuth is from the School of Natural and Complementary Medicine at Southern Cross University in Australia.

36   Personal communication, 9 November 2006.

37   Dr P. Josling is from the Garlic Centre in Battle, East Sussex, UK.

38   P. Josling, 'Preventing the common cold with a garlic supplement: A double-blind, placebo-controlled survey', *Advances in Therapy*, 2001, vol. 18, no. 4, pp. 189–193.

39   S. Juan, 'Does garlic ward off the common cold?', *The Register*, 24 November 2006.

40   Drs A. and M. Eidi, and E. Esmaeili are from the Department of Biology, Science & Research at the Islamic Azad University in Tehran, Iran.

41   A. Eidi et al., 'Antidiabetic effect of garlic (*Allium sativum L.*) in normal and streptozotocin-induced diabetic rate', *Phytomedicine*, 2006, vol. 13, nos. 9–10, pp. 624–629.

42   Drs T. Friedman, A. Shalom and M. Westreich are from the Department of Plastic Surgery at the Assaf Harofeh Medical Centre in Zerifin, Israel.

43   T. Friedman et al., 'Self-inflicted garlic burns: Our experience and literature review', *International Journal of Dermatology*, 2006, vol. 45, no. 10, pp. 1161–1163.

44   Dr Charles Walcott is from the Cornell Laboratory of Ornithology in Ithaca, New York.

45   C. Walcott, 'Magnetic maps in pigeons', EXS, 1991, vol. 60, pp. 38–51.

46   Drs Dennis J. Walmsley and W. Epps are from the Department of Human Geography at the Australian National University in Canberra.

47   D. Walmsley and W. Epps, 'Direction-finding in humans: Ability of individuals to orient towards their place of residence', *Perceptual and Motor Skills*, 1987, vol. 64, no. 3, pt. 1, pp. 744–746.

48   S. Juan, 'Do humans have an inbuilt compass?', *Epoch Times*, 8 November 2006, p. 11.

49   S. Juan, 'Do humans have a compass in their nose?', *The Register*, 17 November 2006.

Chapter 5: The Ears

1   Drs L. Pelz and B. Stein are from the Medical Branch of the University of Rostock in Germany.

2   L. Pelz and B. Stein, 'Clinical assessment of ear size in children and adolescents', *Padiatrie und Grenzgebiete*, 1990, vol. 29, no. 3, pp. 229–235.

3   Dr James Heathcote is a general practitioner from Kent in the UK.

4   J. Heathcote, 'Why do old men have big ears?', *British Medical Journal*, 1995, vol. 311, p. 1668.

5   Dr Yashhiro Asai is a physician at the Futanazu Clinic in Misaki in Japan.

6   Y. Asai, M. Yoshimura, N. Nago and T. Yamada, 'Correlation of ear length with age in Japan', *British Medical Journal*, 1996, vol. 312, p. 582.

7   Dr V.F. Ferrario, C. Sforza, V. Ciusa, G. Serrao and G. Tartaglia are from the Functional Anatomy Research Centre at the University of Milan in Italy.

8   V. Ferrario et al., 'Morphometry of the normal human ear: A cross-sectional study from adolescence to mid-adulthood', *Journal of Craniofacial Genetics and Developmental Biology*, 1999, vol. 19, no. 4, pp. 226–233.

9   M. Woods, 'As we age and shrink, our ears grow on', *Post Gazette* (Pittsburgh), 4 November 2003, pp. 1–2.

10   S. Juan, 'Do our ears grow longer with age?', *The Register*, 26 May 2006.

11   S. Juan, 'Yes, your ears are growing', *National Post*, 26 June 2006, pp. 1–2.

12   Dr Steven Mithen is a professor of early prehistory at the University of Reading in the UK.

13   S. Mithen, *The Singing Neanderthals: The Origins of Music, Language, Mind and Body*, Weidenfeld & Nicholson, London, 2005, pp. 172–173.

14   S. Mithen, 'Moved by the music', *New Scientist*, 16 July 2005, pp. 46–47.

15   C. Sutton, *How Did They Do That?*, Quill, New York, 1985, pp. 261–264.

16   Dr P.D. Shearer is from the St Jude Children's Research Hospital in Memphis, Tennessee.

17   P. Shearer, 'The deafness of Beethoven: An audiologic and medical overview', *American Journal of Otology*, 1990, vol. 11, no. 5, pp. 370–374.

18   Drs C.S. Karmody and E.S. Bachor are from the Tufts University School of Medicine in Boston.

19 C. Karmody and E. Bachor, 'The deafness of Ludwig van Beethoven: An immunopathy', *Otology and Neurotology*, 2005, vol. 26, no. 4, pp. 809–814.
20 Dr R.H. Ratnasuriya is a psychologist at the Bethlem Royal and Maudsley Hospital in London.
21 R. Ratnasuriya, 'Joan of Arc, creative psychopath: Is there another explanation?', *Journal of the Royal Society of Medicine*, 1986, vol. 79, pp. 234–235.
22 Dr D.A. Moore is the medical services director of the Scottish and Newcastle Breweries in Edinburgh, Scotland.
23 D. Moore, 'Response to "Joan of Arc, creative psychopath: Is there another explanation?"', *Journal of the Royal Society of Medicine*, 1986, vol. 79, p. 560.
24 Dr Rudolph Bell is a historian at Rutgers University in Chicago.
25 R. Bell, *Holy Anorexia*, University of Chicago Press, Chicago, 1985.
26 Drs E. Foote-Smith and L. Bayne are from the Department of Neurology at the University of California in San Francisco.
27 E. Foote-Smith and L. Bayne, 'Joan of Arc', *Epilepsia*, 1991, vol. 32, no. 6, pp. 810–815.
28 Dr Maggie Phillips is a psychologist in Oakland, California.
29 M. Phillips, 'Joan of Arc meets Mary Poppins: Maternal re-nurturing approaches with male patients in Ego-State Therapy', *American Journal of Clinical Hypnosis*, 2004, vol. 47, no. 1, pp. 3–12.
30 S. Juan, 'Joan of Arc's secret', *National Post*, 24 April 2006, pp. 1–2.
31 D. Fucci, L. Petrosino, B. Hallowell, L. Andra and C. Wilcox, 'Magnitude estimation scaling of annoyance in response to rock music: Effects of sex and listeners' preference', *Perceptual & Motor Skills*, 1997, vol. 84, no. 2, pp. 663–670.
32 J. Kellaris and R. Kent, 'An exploratory investigation of responses elicited by music varying in tempo, tonality and texture', *Journal of Consumer Psychology*, 1993, vol. 2, no. 4, pp. 381–401.
33 Dr John Manning is from the School of Biological Sciences at the University of Liverpool in the UK.
34 Reuters, 'Ears a way to show men's wretched moods', *Daily Telegraph* (Sydney), 24 July 1997, p. 26.
35 S. Juan, 'The Odd Body: Can you judge a person by their ears?', *Epoch Times*, 18 October 2006, p. 16.
36 S. Juan, 'Can you judge someone's personality by the shape of their ears?', *The Register*, 27 October 2006.
37 S. Juan, 'Great moments in human research', *The Register*, 27 January 2007.
38 S. Juan, 'Great moments in human research', *The Register*, 3 February 2007.

Chapter 6: The Mouth

1 Drs L.E. Cuevas and C.A. Hart are from the Department of Tropical Paediatrics at the Liverpool School of Tropical Medicine at the University of Liverpool, UK.
2 L. Cuevas and C. Hart, 'Chemoprophylaxis of bacterial meningitis', *Journal of Antimicrobial Chemotherapy*, 1993, vol. 31, suppl. B, pp. 79–91.
3 Drs Rosemary Hallett, L.A. Haapanen and S.S. Teuber are from the School of Medicine at the University of California in Davis.

4   R. Hallett et al., 'Food allergies and kissing', *New England Journal of Medicine*, 2002, vol. 346, no. 23, pp. 1833–1834.

5   Dr H. Kimata is from the Department of Allergy at Satou Hospital in Osaka.

6   H. Kimata, 'Kissing selectively decreases allergen-specific IgE production in atopic patients', *Journal of Psychosomatic Research*, 2006, vol. 60, no. 5, pp. 545–547.

7   Drs K. Floyd, J. Boren, A. Hannawa, B. McEwan and A. Veksler are from the Department of Communication at Arizona State University.

8   K. Floyd et al., 'Kissing in marital and cohabiting relationships: Effects on blood lipids, stress, and relationship satisfaction', *Western Journal of Communication*, 2009, vol. 73, no. 2, pp. 113–133.

9   R. Gordon (ed.), *Ethnologue: Languages of the World* (15th edn), Summer Institute of Linguistics, Dallas, 2005, p. 122.

10 Dr Michael Cole is a professor of psychology at the University of California in San Diego.

11 M. Cole and S. Cole, *The Development of Children* (4th edn), Worth Publishing, San Francisco, 2001.

12 S. Juan, 'Umm ...', *The Register*, 6 May 2006.

13 E. Tan, A. Ciger and T. Zileli, 'Whistling epilepsy: A case report', *Clinical Electroencephalography*, 1990, vol. 21, no. 2, pp. 110–111.

14 T. Murray, 'Dr Samuel Johnson's movement disorder', *British Medical Journal*, 1979, vol. 1 (6178), pp. 1610–1614.

15 S. Juan, 'What's happened to whistling?', *The Register*, 2 June 2006.

16 S. Juan, 'Not just whistling Dixie', *National Post*, 25 July 2006.

17 S. Juan, 'What are tag questions?', *The Register*, 2 June 2006.

18 S. Juan, 'How do I taste things?', *The Register*, 4 August 2006.

19 Drs T. Manrique, I. Moron, M. Ballesteros, R. Guerrero and M. Gallo are from the Institute of Neurosciences F. Oloriz of the Department of Experimental Psychology and Physiology of Behaviour at the University of Granada in Spain.

20 T. Manrique et al., 'Hippocampus, ageing and taste memories', *Chemical Senses*, 2007, vol. 32, no. 1, pp. 111–117.

21 C. Wysocki, 'Do people lose their senses of smell and taste as they age?', *Scientific American*, June 2003, p. 107.

22 J. Varasdi, *Myth Information: More than 590 Popular Misconceptions, Fallacies and Misbeliefs Explained*, Ballantine, New York, 1996, p. 234.

23 S. Juan, 'Is it possible to swallow while standing on your head?', *The Register*, 11 August 2006.

24 S. Juan, 'Great moments in human research', *The Register*, 27 January 2007.

25 S. Juan, *The Odd Body: Mysteries of Our Weird and Wonderful Bodies Explained*, HarperCollinsAustralia, Sydney, 1995, p. 97.

26 S. Juan, 'Why are we not irritated by the volume of our own voice?', *The Register*, 22 September 2006.

27 Drs Robert Krauss, R. Freyberg and E. Morsella are from the Department of Psychology at Columbia University.

28 R. Krauss et al., 'Inferring speakers' physical attributes from their voices', *Journal of Experimental Social Psychology*, 2002, vol. 38, pp. 618–625.

29 Drs Susan Hughes, M. Harrison and G. Gallop are from the Department of Psychology at the State University of New York in Albany.

30 S. Hughes et al., 'The sound of symmetry: Voice as a marker of developmental instability', *Evolution and Human Behaviour*, 2002, vol. 23, pp. 173–178.

31 S. Hughes, F. Dispenza and G. Gallop, 'Ratings of voice attractiveness predict sexual behaviour and body configuration', *Evolution and Human Behaviour*, 2004, vol. 25, pp. 295–304.

32 S. Juan, 'What can you learn from the sound of someone's voice?', *The Register*, 6 October 2006.

33 Dr William Sharp is from the Department of Psychology at the University of Mississippi in Oxford.

34 W. Sharp, C. Sherman and A. Gross, 'Selective mutism and anxiety: A review of the current conceptualization of the disorder', *Journal of Anxiety Disorders*, 2006, vol. 21, no. 4, pp. 568–579.

35 S. Juan, 'What is selective mutism?', *The Register*, 9 December 2006.

36 S. Juan, 'Why do we open our mouths to yawn properly?', *The Register*, 13 January 2007.

37 Dr G. Hauser is from the Histological and Embryological Institute of the University of Wien in Germany.

38 G. Hauser, A. Daponte and M. Roberts, 'Palatal rugae', *Journal of Anatomy*, 1989, vol. 165, pp. 237–249.

39 S. Juan, 'Great moments in human research', *The Register*, 3 February 2007.

Chapter 7: The Skin

1 Dr Jonathan Kantor is from the Department of Dermatology at the University of Pennsylvania Medical Centre in Philadelphia.

2 J. Kantor, *Medical Encyclopedia: Body Lice*, US National Library of Medicine and the National Institutes of Health, Bethesda, Maryland, 19 May 2005.

3 S. Juan, 'What lies without: Life on the human body', *The Register*, 2 June 2006.

4 Drs N. Agarwal, A. Kriplani, A. Gupta and N. Bhatia are from the Department of Obstetrics and Gynaecology at the All India Institute of Medical Sciences in New Delhi.

5 N. Agarwal et al., 'Management of gigantomastia complicating pregnancy. A case report', *Journal of Reproductive Medicine*, 2002, vol. 47, no. 10, pp. 871–874.

6 S. Juan, 'Is it true that a woman's breasts can grow enormously overnight?', *The Register*, 13 January 2006.

7 Drs J.M. Wu, A. Mamelak, R. Nussbaum and P. McElgunn are from the School of Medicine at the University of Sheffield.

8 J. Wu et al., 'Botulinum toxin A in the treatment of chromhidrosis', *Dermatological Surgery*, 2005, vol. 31, no. 8 (pt 1), pp. 963–965.

9 Dr Stephen Amon is head of the Infant Botulism Prevention Program at the California Department of Health Sciences in Sacramento.

10 Dr A. Boer is a dermatologist in Hamburg, Germany.

11 A. Boer, 'Patterns histopathologic of Fox-Fordyce disease', *American Journal of Dermatopathology*, 2004, vol. 26, no. 6, pp. 482–492.

12   M. Sims, *Adam's Navel: A Natural and Cultural History of the Human Body*, Allen Lane, London, 2003, pp. 13, 28, 44.

13   S. Juan, 'Why don't humans moult?', *The Register*, 26 June 2006.

14   M. Rogers, 'They've got a hide!', *Medical Observer*, (Sydney) 8 November 2002, p. 56.

15   S. Juan, 'Was human skin really used in book binding?', *The Register*, 4 August 2006.

16   Dr Peter Cave is a philosopher at the Open University, City University of London.

17   P. Cave, 'Birthday special: John Stuart Mill', *Philosophy Now* (London), May–June 2006, pp. 26–29.

18   S. Juan, 'Making rash judgment on philosopher's death', *National Post*, 7 November 2006, pp. 1–2.

19   Dr Sarah-Jayne Blakemore is from the Institute of Cognitive Neuroscience at the University College in London.

20   S. Blakemore, 'Deluding the motor system', *Consciousness and Cognition*, 2003, vol. 12, issue 4, pp. 647–655.

21   Dr D.S. Bennett is from the MCP Hahnemann University in Philadelphia.

22   D. Bennett, M. Bendersky and M. Lewis, 'Facial expressivity at 4 months: A context by expression analysis', *Infancy*, 2002, vol. 3, no. 1, pp. 97–113.

23   Dr M. Blagrove is from the Department of Psychology at the University of Wales in Swansea.

24   M. Blagrove, S. Blakemore and B. Thayer, 'The ability to self-tickle following Rapid Eye Movement sleep dreaming', *Consciousness and Cognition*, 2006, vol. 15, no. 2, pp. 285–294.

25   S. Juan, 'What is the purpose of tickling?', *The Register*, 1 September 2006.

26   J. Varasdi, *Myth Information*, p. 218.

27   S. Juan, 'Why do your hands turn white when you wash the dishes?', *The Register*, 1 September 2006.

28   Dr R. James Swanson is from the Faculty of Biological Sciences at Old Dominion University in Norfolk, Virginia.

29   Personal communication, 2 February 2006.

30   S. Juan, 'Why do you sometimes shiver when you wee?', *The Register*, 1 September 2006.

31   Dr Randolph Morgan is director of the Insectarium at the Cincinnati Zoo in Ohio.

32   D. Feldman, *What are Hyenas Laughing at, Anyway?*, G.P. Putnam, New York, 1995, p. 179.

33   New Jersey Mosquito Control Association, *FAQs on Mosquitoes*, Rutgers University, New Brunswick, New Jersey, 24 October 2006.

34   Dr James Logan is from the Rothamsted Research Institute in Herfordshire, UK, and Professor Jenny Mordue is from the University of Aberdeen in Scotland.

35   Biotechnology and Biological Sciences Research Council, 'Biting back at flies', *BBSRC Business* (London), January 2005, pp. 14–15.

36   C. Ray, 'Mosquitoes and genes', *New York Times*, 16 September 2003, p. D4.

37   New Jersey Mosquito Control Association, *FAQs on Mosquitoes*, Rutgers University, New Brunswick, New Jersey, 24 October 2006.

38 Dr Steven Schutz is from the Mosquito Control Research Laboratory of the University of California at Davis.

39 D. Feldman, *What are Hyenas Laughing at, Anyway?*, pp. 177–178.

40 Dr Leslie Saul-Gershenz is the director of Entomology at the San Francisco Zoological Society.

41 D. Feldman, *What are Hyenas Laughing at, Anyway?*, p. 178.

42 J. Walters and A. O'Donoghue, 'Mozzie attack', *New Scientist*, 12 April 1997, p. 65.

43 J. Richfield and Y. Van Bergen, 'Biting back', *New Scientist*, 29 April 2006, p. 73.

44 S. Juan, 'Why are some people more attractive to mosquitoes?', *The Register*, 10 November 2006.

45 S. Juan, 'How can objects in the same room be different temperatures?', *The Register*, 10 November 2006.

46 W. Fitzpatrick, *An Open Letter to the World About Colloidal Silver*, Argyria Information Website, 13 November 2006.

47 British Broadcasting Commission, 'True-blue bids for Senate', *BBC News* (London), 2 October 2002.

48 S. Juan, 'How the "true blue" political maverick gave the Senate to the donkeys', *The Register*, 9 December 2006.

49 Drs Alan Ashworth, B. Howard, H. Panchal and A. McCarthy are from the Breakthrough Breast Cancer Research Centre in London.

50 A. Ashworth et al., 'Identification of the scaramanga gene implicates Neuregulin 3 in mammary gland specification', *Genes and Development*, 2005, vol. 17, no. 17, pp. 2078–2090.

51 Drs D.M. Conde, E. Kashimoto, R. Torresan and M. Alvarenga are from the Department of Gynecology and Obstetrics at the Hospital Estadual Sumare and the Universidade Estadual de Campinas in Sumara, Brazil.

52 D. Conde et al., 'Pseudomamma on the foot: An unusual presentation of supernumerary breast tissue', *Dermatology Online Journal*, 2006, vol. 12, no. 4, p. 7.

53 S. Juan, 'Why do some people have three nipples?', *The Register*, 13 January 2006.

54 Dr C.R. Goding is from the Marie Curie Research Institute in Surrey, UK.

55 C. Goding, 'Melanocytes: The new black', *International Journal of Biochemistry and Cell Biology*, 2007, vol. 39, no. 2, pp. 275–279.

56 Personal communication, 22 December 2006.

57 S. Juan, 'Why do we like to scratch a wound when it's healing?', *The Register*, 10 February 2007.

58 Drs Hui-Jun Ma, G. Zhao, F. Shi and Y. Wang are from the Department of Dermatology at the Air Force Hospital of the Chinese People's Liberation Army in Beijing.

59 H. Ma et al., 'Eruptive cherry angiomas associated with vitiligo: Provoked by topical nitrogen mustard?', *Journal of Dermatology*, 2006, vol. 33, no. 12, p. 877.

60 EnergyAustralia, *Shower Timers Help Families Become Energy Efficient*, EnergyAustralia, Sydney, October 2006.

61 S. Juan, 'Do we really need a daily shower or bath to stay healthy?', *The Register*, 17 February 2007.

62 S. Juan, 'Great moments in human research', *The Register*, 27 January 2007.

63 S. Juan, 'Great moments in human research', *The Register*, 3 February 2007.

Chapter 8: The Hair & Nails

1   M. Sims, *Adam's Navel: A Natural and Cultural History of the Human Form*, Viking, New York, 2003, pp. 25–27.

2   S. Juan, 'Why we are not naked even in the womb', *The Register*, 7 May 2006.

3   M. Symons, *Why Girls Can't Throw: … and Other Questions You Always Wanted Answered*, Harper Paperbacks, New York, 2006, pp. 172–173.

4   S. Juan, 'Will eating crusts make your hair grow curly?', *The Register,* 4 August 2006.

5   S. Juan, 'Keeping it in the family', *National Post*, 20 November 2006, pp. 1–2.

6   H. Sustaita, *A Close Look at the Properties of Hair and Scalp*, Houston Community College — Northwest, Houston, 2006, pp. 7–9.

7   S. Juan, 'Why isn't pubic hair the same colour as hair on your head?', *The Register*, 15 September 2006.

8   Dr John O'Connor is head of the School of Physical Science and Mathematics at the University of Newcastle in Australia.

9   Personal communication, 12 August 2006.

10 S. Juan, 'Can your hair turn white as a result of shock?', *The Register*, 29 September 2006.

11 Dr John Mason is a trichologist from Royston near Barnsley in South Yorkshire, UK.

12 J. Mason, 'The role of the trichologist', *Clinical and Experimental Dermatology*, 2002, vol. 27, no. 5, pp. 422–425.

13 Personal communication, 12 November 2006.

14 S. Juan, 'Why is the human face hairless?', *The Register*, 20 January 2007.

15 P. Abrahams, *How the Body Works*, Amber Books, London, 2007. p. 415.

16 Dr Alan Greene is a clinical professor at the School of Medicine at Stanford University and chief medical officer of ADAM, a medical information organisation.

17 Dr Robert Baran is from the Nail Disease Centre in Cannes, France.

18 J. Brody, 'Fingernails can reveal much about habits and health', *New York Times*, 22 January 1990, pp. B3–B4.

Chapter 9: The Skeleton, Bones & Teeth

1   Committee on Sports Medicine and Fitness and Council on School Health, American Academy of Pediatrics, 'Active healthy living: Prevention of childhood obesity through increased physical activity', *Pediatrics*, 2006, vol. 171, no. 5, pp. 1834–1842.

2   National Association for Sport and Physical Education, *Active Start: A Statement of Physical Activity Guidelines for Children Birth to Five Years,* National Association for Sport and Physical Education, Reston, Virginia, 2001.

3   Centre for Community Child Health, Royal Children's Hospital, Melbourne, 'Leaping early in life', *Community Paediatric Review*, 2006, vol. 15, no. 4, pp. 4–5.

4   H. Klawans, *Strange Behaviour: Tales of Evolutionary Neurology*, Norton, New York, 2000, pp. 37–38.

5   Drs A.K. Tan and C.B. Tan are from the Department of Neurology at Tan Tock Seng Hospital in Singapore.

6   A. Tan and C. Tan, 'The syndrome of painful legs and moving toes — a case report', *Singapore Medical Journal*, 1996, vol. 37, no. 4, pp. 446–447.

7   M. Sims, *Adam's Navel: A Natural and Cultural History of the Human Form*, Penguin Books Australia, Melbourne, 2004, pp. 292–295.

8   Drs S.S. Campbell and P.J. Murphy are from the Laboratory of Human Chronobiology in the Department of Psychiatry at Cornell University Medical College in White Plains, New York.

9   S. Campbell and P. Murphy, 'Extraocular circadian phototransduction in humans', *Science*, 1998, vol. 279 (5349), pp. 333–334.

10   K. Hopkin, 'Clock setting', *Scientific American*, April 1998, pp. 20–22.

11   Dr S. Ooki is from the Department of Health Science of Ishikawa Prefectural Nursing University in Kahoku, Japan.

12   S. Ooki, 'Genetic and environmental influences on finger-sucking and nail-biting in Japanese twin children', *Twin Research and Human Genetics*, 2005, vol. 8, no. 4, pp. 320–327.

13   Drs B. Mangweth, A. Hausmann, C. Danzl, T. Walch, C. Rupp, W. Biebl, J. Hudson and H. Pope are from the Department of Psychiatry at the Innsbruck University Hospital in Austria.

14   B. Mangweth et al., 'Childhood body-focused behaviours and social behaviours as risk factors of eating disorders', *Psychotherapy and Psychosomatics*, 2005, vol. 74, no. 4, pp. 247–253.

15   Drs S. Yassaei, M. Rafieian and R. Ghafari are from the Shahid Sadoughi University of Medical Sciences and Health Services in Yazd, Iran.

16   S. Yassaei et al., 'Abnormal oral habits in the children of war veterans', *Journal of Clinical Pediatric Dentistry*, 2005, vol. 29, no. 3, pp. 189–192.

17   Drs P.G. Hepper, D. Wells and C. Lynch are from the School of Psychology at Queen's University in Belfast, Northern Ireland.

18   P. Hepper et al., 'Prenatal thumb sucking is related to postnatal handedness', *Neuropsychologia*, 2005, vol. 43, no. 3, pp. 313–315.

19   S. Juan, 'Does thumb-sucking run in families?', *The Register*, 4 August 2006.

20   S. Juan, 'Keeping it in the family', *National Post*, 20 November 2006, pp. 1–2.

21   C. Frey, *The Female Fleet of Foot*, American Orthopaedic Foot and Ankle Society, Rosemont, Illinois, 20 July 2006.

22   S. Juan, 'Why do women have smaller feet?', *Epoch Times*, 27 July 2006, p. 8.

23   S. Juan, 'Why do women have smaller feet?', *The Register*, 11 August 2006.

24   S. Juan, 'Great moments in human research', *The Register*, 27 January 2007.

25   S. Juan, 'Great moments in human research', *The Register*, 3 February 2007.

26   B. Hee, 'Water baby', *New Scientist*, 28 June 1997, p. 65.

27   S. Juan, 'What makes a good swimmer?', *The Register*, 25 August 2006.

28   Dr Kevin Beck is a psychologist from Human Kinetics Inc in Champaign, Illinois.

29   K. Beck, *Choosing Optimal Stride Length*, Human Kinetics Inc, Champaign, Illinois, 28 August 2006.

30   Dr R. McNeill Alexander is from the School of Biology at the University of Leeds in the UK.

31   R. McNeill Alexander, 'Energetics and optimisation of human walking and running: The 2000 Raymond Pearl memorial lecture', *American Journal of Human Biology*, 2002, vol. 14, no. 5, pp. 641–648.

32   A. Ward-Smith, 'The bioenergetics of optimal performances in middle distance and long-distance track running', *Biomechanics*, 1999, vol. 32, no. 5, pp. 461–465.

33   A. Ward-Smith, 'Energy conversion strategies during 100 m sprinting', *Journal of Sports Science*, 2001, vol. 19, no. 9, pp. 701–710.

34   Dr William Sellars is from the Department of Human Sciences at Loughborough University in the UK.

35   W. Sellars, G. Cain, W. Wang and R. Crompton, 'Stride lengths, speed and energy costs in walking of *Australopithecus afraensis*: Using evolutionary robotics to predict locomotion of early human ancestors', *Journal of the Royal Society Interface*, 2005, vol. 2, pp. 431–441.

36   S. Juan, 'Is there a speed or stride where running is more efficient?', *The Register*, 22 September 2006.

37   J. Varasdi, *Myth Information*, 1996, p. 238.

38   Dr A. Bazile, N. Bissada, R. Nair and B. Siegel are from the Department of Periodontics at Case Western Reserve University in Cleveland, Ohio.

39   A. Bazile et al., 'Periodontal assessment of patients undergoing angioplasty for treatment of coronary artery disease', *Journal of Periodontology*, 2002, vol. 73, no. 6, pp. 631–636.

40   Dr Barbara Taylor is head of periodontics at the Sydney Dental Hospital.

41   B. Taylor, G. Tofler, H. Carey, M. Morel-Kopp, S. Philcox, T. Carter, M. Elliott, A. Kull, C. Ward and K. Schenck, 'Full mouth tooth extraction lowers systemic inflammatory and thrombotic markers of cardiovascular risk', *Journal of Dental Research*, 2006, vol. 85, no. 1, pp. 74–78.

42   Dr Mark Herzberg is a professor of preventive sciences at the University of Minnesota in Minneapolis.

43   R. Smith, 'Can flossing prevent a heart attack?', *Health*, May 1998, p. 136.

44   S. Juan, 'Can flossing your teeth prevent a heart attack?', *The Register*, 29 September 2006.

45   Dr Holly Muggleston is from the School of Health and Applied Sciences at Southern Cross University in New South Wales.

46   Personal communication, 10 November 2006.

47   S. Juan, 'Will sweets really rot your teeth?', *The Register*, 24 November 2006.

Chapter 10: The Heart, Blood & Lungs

1   Dr Richard Jonas is from the Children's National Heart Institute in Washington, DC.

2   J. Varasdi, *Myth Information*, pp. 123–124.

3   J. Varasdi, *Myth Information*, pp. 122–123.

4   E. Widmaier, *Why Geese Don't Get Obese (And We Do): How Evolution's Strategies for Survival Affect Our Everyday Lives*, W.H. Freeman, San Francisco, 1998, p. 178.

5   Drs A. Pelliccia, B. Maron, A. Spataro, M. Proschan and P. Spirito are from the Department of Medicine at the Comitato Olimpico Nazionale Italiano in Rome.

6   A. Pelliccia et al., 'The upper limit of physiologic cardiac hypertrophy in highly trained elite athletes', *New England Journal of Medicine*, 1991, vol. 324, no. 25, pp. 1812–1813.

7   Dr Alfred Goldberg is a professor of cell biology at Harvard Medical School.

8   J. Shaw, 'The deadliest sin', *Harvard Magazine*, March–April 2004, pp. 36–43, 98–99.

9   C. Petit, 'What causes arteries to harden', *San Francisco Chronicle*, 17 January 1995, p. 2.

10   National Heart Lung and Blood Institute, *What is Peripheral Arterial Disease?*, US Department of Health and Human Services, Washington, DC, 2 January 2007.

11   A. Ciocco, 'On the interdependence of the length of life of husband and wife', *Human Biology*, 1941, vol. 13, no. 4, pp. 505–525.

12   W. Corliss, *Biological Anomalies II*, Sourcebook Project, Glen Arm, Maryland, 1993, pp. 250–251.

13   M. Higgins, J. Keller, F. Moore, L. Ostrander, H. Metzner and L. Stock, 'Studies of blood pressure in Tecumseh, Michigan. I. Blood pressure in young people and its relationship to personal and familial characteristics and complications of pregnancy in mothers', *American Journal of Epidemiology*, 1980, vol. 11, no. 2, pp. 142–145.

14   S. Juan, 'Is heart rate correlated with birth order?', *The Register*, 7 July 2006.

15   Dr Sally Edwards is the CEO of Heart Zones USA of Sacramento, California.

16   Personal communication, 10 December 2006.

17   S. Juan, 'How does the heart differ from other mechanical pumps?', *The Register*, 23 December 2006.

18   M. Goldwyn, *How a Fly Walks Upside Down ... and Other Curious Facts*, Wings, Atlanta, 1995, p. 40.

19   S. Juan, 'What makes a wound stop bleeding?', *The Register*, 26 June 2006.

20   *Fast Facts*, National Hemophilia Foundation, New York, 25 May 2006.

21   J. Varasdi, *Myth Information*, p. 125.

22   S. Juan, 'What happened to hemophiliacs before blood supplies were safe?', *The Register*, 6 June 2006.

23   Drs K. Kasirajan, R. Milner and E. Chaikof are from the School of Medicine at Emory University in Atlanta, Georgia.

24   K. Kasirajan et al., 'Combination therapies for deep venous thrombosis', *Seminars in Vascular Surgery*, 2006, vol. 19, no. 2, pp. 116–121.

25   S. Juan, 'What is deep vein thrombosis?', *The Register*, 14 July 2006.

26   American Red Cross, New England Region, *Frequently Asked Questions*, Boston, 2 August 2006.

27   S. Juan, 'What conditions disqualify you from donating blood?', *The Register*, 25 August 2006.

28   World Health Organization, *Tuberculosis Fact Sheet*, Geneva, March 2006.

29   Drs G.A. Lammie, R. Hewlett, J. Shoeman and P. Donald are from the Department of Pathology at Cardiff University in Wales.

30   G. Lammie et al., 'Tuberculosis encephalopathy: A reappraisal', *Acta Neuropathologica* (Berlin), 2007, vol. 113, no. 3, pp. 227–234, Epub 14 December 2006.

31 J. Varasdi, *Myth Information*, 1996, p. 246.

32 S. Juan, 'Whatever happened to tuberculosis?', *The Register*, 3 February 2007.

Chapter 11: The Stomach & Intestines

1 Dr Peter Osin is from the Royal Marsden Hospital in London.

2 'Why the fat lady sings — there may be a medical reason opera singers tend to be heavy', *Evening Standard* (London), 28 October 2005.

3 Drs C.W. Thorpe, S. Cala, J. Chapman and P. Davis are from the National Voice Centre at the University of Sydney.

4 C. Thorpe et al., 'Patterns of breath support in projection of the singing voice', *Journal of Voice*, 2001, vol. 15, no. 1, pp. 86–104.

5 P. Galek, 'Size matters', *New Scientist*, 15 June 2002, p. 63.

6 S. Juan, 'Why are opera singers fat?', *The Register*, 26 June 2006.

7 S. Juan, 'Weighing in on opera singers' physiques', *National Post*, 2 October 2006, pp. 1–2.

8 Centres for Disease Control and Prevention, *Tips for Preventing Heat-Related Illness*, Atlanta, 25 July 2006.

9 S. Juan, 'What fluids should you drink when it's hot?', *The Register*, 18 August 2006.

10 Dr Andrew Lloyd is an infectious disease physician in Sydney.

11 Personal communication, 22 July 2006.

12 Dr Holly Muggleston is from the School of Health and Applied Sciences at Southern Cross University in New South Wales.

13 Personal communication, 24 July 2006.

14 S. Juan, 'Do you feed a cold and starve a fever?', *The Register*, 18 August 2006.

15 Dr Shawn Somerset is from the School of Public Health at Griffith University in Brisbane, Queensland.

16 Personal communication, 9 August 2006.

17 S. Juan, 'Will eating spinach make me strong?', *The Register*, 18 August 2006.

18 S. Juan, 'Why don't we suffer from *E. coli* all the time?', *The Register*, 8 September 2006.

19 H. Gleitman, A. Fridlund and D. Reisberg, *Psychology* (6th edn), W.W. Norton, New York, 2003.

20 S. Juan, 'Why do you sometimes lose bowel function when scared?', *The Register*, 20 October 2006.

21 S. Juan, 'Great moments in human research', *The Register*, 27 January 2007.

22 S. Juan, 'Great moments in human research', *The Register*, 3 February 2007.

23 American Liver Society, *Gallstones*, Nashville, Tennessee, 4 December 2006.

24 Dr Terry Bolin is a gastroenterologist at the Gut Foundation Institute at the Prince of Wales Hospital and the University of New South Wales in Sydney.

25 Personal communication, 4 December 2006.

26 R. Kazmierski, 'Primary adenocarcinoma of the gallbladder with intramural calcification', *American Journal of Surgery*, 1951, vol. 82, pp. 248–250.

27 Drs Tsung-Chun Lee, K. Liu, I. Lai and H. Wang are from the National Taiwan University Hospital in Taiwan.

28 T. Lee et al., 'Diagnosing porcelain gallbladder', *American Journal of Medicine*, 2005, vol. 118, no. 10, pp. 1171–1172.

29 S. Juan, 'Are women who are forty, fat and fair more likely to get gallstones?',
     The Register, 20 January 2006.

30 J. Arnold, 'Scientific sleuths track the origins of tapeworms in humans', *ARS US
     Department of Agriculture*, 23 October 2006, p. 1.

31 J. Varasdi, *Myth Information*, pp. 237–238.

32 S. Juan, 'How much damage does a tapeworm do to the human body?',
     The Register, 27 January 2007.

33 T. Lee and D. Hardman, 'The morning before', *New Scientist*, 9 June 2001, p. 65.

34 Drs G. Koren and C. Maltepe are from the Hospital for Sick Children in Toronto,
     Canada.

35 G. Koren and C. Maltepe, 'Preventing recurrence of severe morning sickness',
     Canadian Family Physician, 2006, vol. 52, no. 12, pp. 1545–1546.

36 Drs C. Louik, S. Hernandez-Diaz, M. Werler and A. Mitchell are from the Slone
     Epidemiology Centre at Boston University.

37 C. Louik et al., 'Nausea and vomiting in pregnancy: Maternal characteristics and
     risk factors', *Paediatric and Perinatal Epidemiology*, 2006, vol. 20, no. 4, pp. 270–278.

38 L. Farina, I. Jeffcoate and R. Lucas, 'Dodgy tummy', *New Science*, 6 January 2007,
     p. 57.

39 J. Varasdi, *Myth Information*, 1996, pp. 210–211.

40 Drs I. Ahmed, D. Deakin and S.L. Parsons are from the Department of General
     Surgery at the Queen's Medical Centre in Nottingham, UK.

41 I. Ahmed et al., 'Appendix mass: Do we know how to treat it?', *Annals of the Royal
     College of Surgeons of England*, 2005, vol. 87, no. 3, pp. 191–195.

42 S. Juan, 'Do we still remove the appendix as often as we used to?', *The Register*,
     26 May 2006.

43 Timothy L. Taylor is a science writer from Vancouver, Canada.

44 T. Taylor, *The Prehistory of Sex: Four Million Years of Human Culture*, Bantam, New
     York, 1996.

45 M. Sims, *Adam's Navel: A Natural and Cultural History of the Human Form*, Penguin
     Books Australia, Melbourne, 2004, pp. 278–281.

Chapter 12: Otherwise Inside

1 S. Juan, 'What lies within', *The Register*, 9 June 2006.

2 S. Juan, 'Why seeing is not always believing', *National Post*, 28 September 2006,
    pp. 1–2.

3 S. Juan, 'How long does it take the body to ...?', *Epoch Times*, 19 May 2006, p. 8.

4 S. Juan, 'Ready, steady, grow', *The Register*, 26 May 2006.

5 Dr K. Nakamura is a chemist and Dr Mitsuo Hiramatsu is a photobiologist at the
    Electron Tube Division of the Central Research Laboratory at Hamamatsu
    Photonics in Hamamatsu, Japan.

6 K. Nakamura and M. Hiramatsu, 'Ultra-weak photon emission from human
    hand: Influence of temperature and oxygen concentration on emission', *Journal
    of Photochemistry and Photobiology B: Biology*, 2005, vol. 80, no. 2, pp. 156–160.

7 Drs M.P. Guedj and A. Lev are from the University Hospital of Hadassah in Israel.

8 M. Guedj and A. Lev, '*Situs inversus*: Leave well alone', *Annales Francaises
    d'Anesthesie et de Reanimation*, 2007, vol. 26, no. 3, pp. 265–266.

References

9   Goddard Space Flight Centre, *Ask an Astronaut*, National Aeronautics and Space Administration, Houston, 12 February 2007.

10 Dr Samuel Conway is from Avid Therapeutics in Philadelphia.

11 Personal communication, 8 February 2007.

12 R. Marsden and C. Robertson, 'Major Toms', *New Scientist*, 31 August 1996, p. 65.

13 Drs Richard Jennings and Ellen Baker are from the University of Texas Medical Branch in Galveston.

14 R. Jennings and E. Baker, 'Gynecological and reproductive issues for women in space: A review', *Obstetrical & Gynecological Survey*, 2000, vol. 55, no. 2, pp. 109–116.

15 J. Jones, R. Jennings, R. Pietryzk, N. Ciftcioglu and P. Stepaniak, 'Genitourinary issues during spaceflight: A review', *International Journal of Impotence Research*, 2005, vol. 17 (Suppl. 1), pp. S64–S67.

16 S. Juan, 'What issues are there for women in space?', *The Register*, 28 July 2006.

17 G. Mirkin, *Catch a Cold*, DrMirkin.com, 13 November 2006.

18 Dr Terry Bolin is a gastroenterologist at the Gut Foundation Institute at the Prince of Wales Hospital and the University of New South Wales in Sydney.

19 Personal communication, 12 November 2006.

20 Dr Andrew Lloyd is from the School of Medicine at the University of Newcastle in New South Wales.

21 M. Ham, 'But my mum said …', *Sydney Morning Herald*, 19 January 2006, 'Health & Science', pp. 4–5.

22 Dr Richard Fedorak is from the University of Alberta Hospital in Edmonton, Canada.

23 Personal communication, 3 January 2007.

24 Dr Roshini Rajapaksa is from the New York University School of Medicine.

25 Personal communication, 3 January 2007.

26 J. Richfield, 'No swimming', *New Scientist*, 12 January 2007, p. 57.

27 Drs A.K. Myhre, K. Berntzen and D. Bratlid are from the Department of Laboratory Medicine, Children's and Women's Health at the Norwegian University of Science and Technology in Trondheim.

28 A. Myhre et al., 'Genital anatomy in non-abused preschool girls', *Acta Paediatrica*, 2003, vol. 92, no. 12, pp. 1453–1462.

29 S. Juan, 'What is the use of the hymen?', *The Register*, 11 August 2006.

30 M. Foster, 'The boob-onic plague', *Weekly World News*, 26 August 2002, pp. 42–43.

31 K. Campbell and M. Lakie, 'Aaaaaah?', *New Scientist*, 15 March 1997, p. 65.

32 S. Juan, 'A blow for babe magnets', *National Post*, 17 April 2006, p. 1.

33 Dr Michael Onken is from the Department of Anatomy at Washington University in St Louis.

34 Personal communication, 30 March 2006.

35 S. Juan, 'A blow for babe magnets', *National Post*, 17 April 2006, p. 1.

36 Stephen Turner is an engineer with the American Society of Heating, Refrigerating and Air-Conditioning Engineers (ASHRAE) in Atlanta, Georgia.

37 Personal communication, 23 August 2006.

38 Personal communication, 24 August 2006.

39 S. Juan, 'Why do some people feel the cold more than others?', *The Register*, 8 September 2006.

40 S. Juan, 'Great moments in human research', *The Register*, 27 January 2007.

41 S. Juan, 'Great moments in human research', *The Register*, 3 February 2007.

42 Drs J.M. Draus, S. Huss, N. Harty, W. Cheadle and G. Larson are from the Department of Surgery at the School of Medicine at the University of Louisville, Kentucky.

43 J. Draus et al., 'Enterocutaneous fistula: Are treatments improving?', *Surgery*, 2006, vol. 140, no. 4, pp. 570–578.

44 S. Juan, 'What is a fistula?', *The Register*, 27 October 2006.

45 S. Juan, 'Does alcohol really keep you warm', *The Register*, 27 October 2006.

46 Dr Gerhard Gmel is from the Alcohol Treatment Centre at Lausanne University Hospital, Switzerland, and Dr Jurgen Rehm is from the Department of Public Health Sciences at the University of Toronto, Canada.

47 G. Gmel and J. Rehm, 'Harmful alcohol use', *Alcohol Research and Health*, 2003, vol. 27, no. 1, pp. 52–62.

Chapter 13: Behaviour

1 S. Juan, 'Three orders of the mind', *Epoch Times*, 20 December 2006, p. 9.

2 R. Campbell, *Campbell's Psychiatric Dictionary*, Oxford University Press, New York, 2004, pp. 538–539, 541.

3 S. Juan, 'Keeping your psychos straight', *National Post*, 15 May 2006, pp. 1–2.

4 Dr John Gartner is a clinical psychologist at Johns Hopkins University School of Medicine.

5 J. Gartner, 'Dark Minds', *Psychology Today*, September–October 2009, pp. 37–38.

6 Dr Patrick Leman is from the Royal Holloway College of the University of London.

7 British Psychological Society, *Who shot the president?*, London, 18 March 2003.

8 P. Leman, 'The born conspiracy', *New Scientist*, 14 July 2007, pp. 35–37.

9 Dr Robert Burns was formerly the Director of the Institute for Human Development at the University of Seattle, Washington.

10 M. Taylor, 'Why we doodle', *In Health*, March–April 1991, pp. 30–33.

11 British Broadcasting Commission, '"Blair" doodles amuse Number 10', *BBC News* (London), 30 January 2005.

12 'Sifting through the mystery of JFK's doodles', *Associated Press* (New York), 23 November 2004.

13 N. Nelson, *The Doodle Dictionary*, Doubleday, New York, 1992.

14 S. Juan, 'Why do we doodle?', *The Register*, 13 October 2006.

15 E. Dykeman, 'Doodling as a memory boost', *Elder Care Express Newsletter* (East Lansing, Michigan), 5 March 2009.

16 Dr Jackie Andrade is from the School of Psychology at the University of Plymouth in the UK.

17 J. Andrade, 'What does doodling do?', *Applied Cognitive Psychology*, 2009, vol. 23, no. 2, Epub 27 February 2009.

18 I. Sample, 'Doodling should be encouraged in boring meetings, claims psychologist', *The Guardian* (London), 27 February 2009, p. 15.

19   Dr Peter Quintieri is from the School of Medicine at Duke University in Durham, North Carolina, and Dr Kenneth Weiss is from the University of Medicine and Dentistry of New Jersey.

20   P. Quintieri and K. Weiss, 'Admissibility of false-confession testimony: Know thy standard', *Journal of the American Academy of Psychiatry and the Law*, 2005, vol. 33, no. 4, pp. 535–538.

21   Drs S.M. Kassin, C. Meissner and R. Norwick are from the Department of Psychology at Williams College in Williamstown, Massachusetts.

22   S. Kassin et al., '"I'd know a false confession if I saw one": A comparative study of college students and police investigators', *Law and Human Behaviour*, 2005, vol. 29, no. 2, pp. 211–227.

23   S. Juan, 'What is a Confessing Sam?', *The Register*, 1 September 2006.

24   J. Leckenby, *Claims About the Power of Subliminal Advertising,* Center for Interactive Advertising, University of Texas, Austin, 6 March 2006.

25   Drs Susan Crawley, C. French and S. Yesson are from the Department of Psychology at Goldsmith College of the University of London.

26   S. Crawley et al., 'Evidence for transliminality from a subliminal card-guessing task', *Perception*, 2002, vol. 31, pp. 887–892.

27   S. Juan, 'Hidden messages and ESP', *National Post,* 27 March 2006, p. 1.

28   P. Marks, 'The lie-detector you'll never know is there', *New Scientist*, 7 January 2006, p. 22.

29   S. Juan, 'Do lie detector tests really work?', *The Register*, 16 June 2006.

30   S. Juan, 'It would be a lie to say that it's a lie-detector', *National Post*, 6 November 2006, p. 1.

31   Drs M.A. Huelsman, J. Piroch and D. Wasieleski are from Coastal Carolina University in Conway, South Carolina.

32   M. Huelsman et al., 'Relation of religiosity with academic dishonesty in a sample of college students', *Psychological Reports*, 2006, vol. 99, no. 3, pp. 739–742.

33   Drs M.K. Johnson, S. Hashtroudi and D.S. Lindsay are from the Department of Psychology at Princeton University, New Jersey.

34   M. Johnson et al., 'Source monitoring', *Psychological Bulletin*, 1993, vol. 114, no. 1, pp. 3–28.

35   Drs S.A. Perkins and E. Turiel are from the Department of Psychology at the University of California at Berkeley.

36   S. Perkins and E. Turiel, 'To lie or not to lie: To whom and under what circumstances', *Child Development*, 2007, vol. 78, no. 2, pp. 609–621.

37   S. Juan, 'How honest are humans', *National Post*, 17 April 2007, pp. 1–2.

38   R. Comer and J. Laird, 'Choosing to suffer as a consequence of expecting to suffer: Why do people do it?', *Journal of Personality and Social Psychology*, 1975, vol. 32, no. 1, pp. 92–101.

39   Drs Rebecca Curtis, Paul Smith and Robert Moore are from the Department of Psychology at Adelphi University in New York.

40   R. Curtis et al., 'Suffering to improve outcomes determined by both chance and skill', *Journal of Social and Clinical Psychology*, 1984, vol. 2, no. 2, pp. 165–173.

41   Drs Roy Baumeister, Jean Twenge and Christopher Nuss are from the Department of Psychology at Case Western Reserve University in Cleveland.

42   R. Baumeister et al., 'Effects of social exclusion on cognitive processes: Anticipated aloneness reduces intelligent thought', *Journal of Personality and Social Psychology*, 2002, vol. 83, no. 4, pp. 817–827.

43   Drs T.B. Kashdan, J. Elhai and B. Frueh are from the Department of Psychology at George Mason University in Virginia.

44   T. Kashdan et al., 'Anhedonia and emotional numbing in combat veterans with PTSD', *Behaviour Research and Therapy*, 2006, vol. 44, no. 3, pp. 457–467.

45   R. Campbell, *Campbell's Psychiatric Dictionary*, Oxford University Press, New York, 2004, p. 40.

46   Dr Ayala Malach-Pines is a psychologist at Ben-Gurion University in Beersheba, Israel.

47   A. Malach-Pines, *Romantic Jealousy: Causes, Symptoms, Cures*, Routledge, London, 1998, p. 102.

48   Dr Salman Akhtar is from the Jefferson Medical College in Philadelphia.

49   S. Akhtar, 'Forgiveness: Origins, dynamics, psychopathology, and technical relevance', *Psychoanalytic Quarterly*, 2002, vol. 71, no. 2, pp. 175–212.

50   Dr David Lotto is from the University of Massachusetts School of Medicine in Boston.

51   D. Lotto, 'The psychohistory of vengeance', *Journal of Psychohistory*, 2006, vol. 34, no. 1, pp. 43–59.

52   Dr Anthony N.G. Clark is a psychologist from Brighton in the UK.

53   Personal communication, 4 March 2005.

54   S. Juan, 'What cultures don't share Western economic values?', *The Register*, 16 June 2006.

55   Dr W. E. Addison is from the Department of Psychology at Eastern Illinois University in Charleston.

56   W. Addison, 'Beardedness as a factor in perceived masculinity', *Perceptual and Motor Skills*, 1989, vol. 68, pp. 921–922.

57   Drs Michael Shannon and C. Patrick Stark are from the Department of Psychology at Western State College in Gunnison, Colorado.

58   M. Shannon and C. Stark, 'The influence of physical appearance on personnel selection', *Social Behaviour and Personality*, 2003, vol. 31, no. 6, pp. 613–623.

59   Drs A.A. de Souza, V.B. Baiao and E. Otta are from the Department of Experimental Psychology at the University of Sao Paulo in Brazil.

60   A. de Souza et al., 'Perception of men's personal qualities and prospect of employment as a function of facial hair', *Psychological Reports*, 2003, vol. 92, no. 1, pp. 201–208.

61   S. Juan, 'Aching heads', *New York Daily News*, 10 May 2005, 'The Odd Body', pp. 1–2.

62   Dr Roy Baumeister, J. Twenge and K. Catanese are from the Case Western Reserve University, Ohio.

63   R. Baumeister et al., 'Social exclusion causes self-defeating behaviour', *Journal of Personality and Social Psychology*, 2002, vol. 83, no. 3, pp. 606–615.

64   J. Briggs, 'I.Q. bruised as ego battered', *Psychology Today*, August 2002, p. 42.

65   S. Juan, 'Women warmer at the core', *National Post*, 6 February 2006, pp. 1–2.

66   Drs O. Egger and M. Rauterberg are from the Swiss Federal Institute of Technology in Lausanne, Switzerland.

67   O. Egger and M. Rauterberg, *Internet Behaviour and Addiction*, Swiss Federal Institute of Technology, Lausanne, 1996.

68   Dr K.S. Young is from the Department of Psychology at the University of Pittsburgh at Bradford, Pennsylvania.

69   K. Young, 'Internet addiction: The emergence of a new clinical disorder', *Cyber Psychology and Behaviour*, 1996, vol. 1, no. 3, pp. 237–244.

70   S. Juan, 'Could you be addicted to the internet?', *The Register*, 22 September 2006.

71   H. Phillips, 'Just can't get enough', *New Scientist*, 26 August 2006, pp. 30–35.

72   Dr Andrew Scholey is a psychopharmacologist at the University of Northumbria in Newcastle, UK.

73   R. Edwards, 'All in the mind?', *New Scientist*, 29 April 2000, p. 19.

74   S. Juan, 'Can you become intoxicated by the power of suggestion?', *The Register*, 8 September 2006.

75   S. Juan, 'Can you become intoxicated by the power of suggestion?', *Epoch Times*, 26 September 2006.

76   Drs M.S. Vos and J.C. de Haes are from the Department of Psychiatry at the Bronovo Hospital in The Hague, The Netherlands.

77   M. Vos and J. Haes, 'Denial in cancer patients, an explorative review', *Psycho-Oncology*, 2007, vol. 16, no. 1, pp. 12–25, Epub 25 July 2006.

78   S. Juan, 'Why are people so often in denial?', *The Register*, 29 September 2006.

79   Dr Marcelo Suarez-Orozco is from the Department of Anthropology at the University of California at San Diego.

80   M. Suarez-Orozco, 'Speaking of the unspeakable: Toward a psychosocial understanding of responses to terror', *Ethos*, 1990, vol. 18, no. 3, pp. 353–383.

81   Dr G.C. Gauchard, J. Muir, C. Touron, L. Benamghar, D. Dehaene, P. Perrin and N. Chau are from the WHO Collaborative Centre in the Faculty of Medicine at the Henri Poincaré University in Nancy, France.

82   G. Gauchard et al., 'Determinants of accident proneness: A case-control study in railway workers', *Occupational Medicine*, 2006, vol. 56, no. 3, pp. 187–190.

83   Professor Ivan Robertson and colleagues are from the Institute of Science and Technology at the University of Manchester, UK.

84   British Broadcasting Commission, 'Identifying the accident prone', *BBC News* (London), 4 January 2001.

85   C. Martin, 'He spent life picking himself up', *The Denver Post*, 23 September 2006.

86   S. Juan, 'What type of person is accident-prone?', *The Register*, 20 October 2006.

87   Dr Dale Larson is from Santa Clara University and Dr R.L. Chastain is from Samuel Merritt College in Oakland, California.

88   D. Larson and R. Chastain, 'Self-concealment: Conceptualisation, measurement, and health implications', *Journal of Social and Clinical Psychology*, 1990, vol. 9, pp. 439–455.

89   Dr J.W. Pennebaker is from the Department of Psychology at the University of Texas in Austin.

90   J. Pennebaker, *Opening Up: The Healing Power of Expression Emotion*, Guilford, New York, 1997.

91   Drs C.E. Hill, C. Gelso and J. Mohr are from the University of Maryland at College Park.

92 C. Hill et al., 'Client concealment and self-presentation in therapy: Comment on Kelly (2000)', *Psychological Bulletin*, 2000, vol. 126, pp. 495–500.

93 Drs Anita Kelly and Jonathan Yip are from the Department of Psychology at the University of Notre Dame in South Bend, Indiana.

94 A. Kelly and J. Yip, 'Is keeping a secret or being a secretive person linked to psychological symptoms?', *Journal of Personality*, 2006, vol. 74, no. 4, pp. 1349–1370.

95 S. Juan, 'Are secretive people more or less healthy?', *The Register*, 1 December 2006.

96 Dr Marie Hartwell-Walker is clinical director of the Adult Outpatient Services Program of the Community Mental Health Centre of Western Massachusetts in Springfield.

97 M. Hartwell-Walker, *Absence Makes the Heart Grow Fonder ... or Does It? The Challenge of Long Distance Relationships*, Psych Central, Newburyport, Massachusetts, 20 April 2006.

98 R. Campbell, *Campbell's Psychiatric Dictionary*, Oxford University Press, New York, 2004, pp. 328, 497.

99 S. Juan, 'Which comes first: Imagination or fantasy?', *The Register*, 19 May 2006.

100 S. Juan, 'Great moments in human research', *The Register*, 27 January 2007.

101 S. Juan, 'Great moments in human research', *The Register*, 3 February 2007.

Chapter 14: Endings

1 T. Cassidy, '3-year-old girl has died 35 times!', *Weekly World News*, 4 August 1998, p. 24.

2 Dr J.B. Stephenson is from the Royal Hospital for Sick Children in Glasgow, Scotland.

3 J. Stephenson, 'Reflex anoxic seizures ("while breath holding"): Nonepileptic vagal attacks', *Archives of Disease in Children*, 1978, vol. 53, no. 3, pp. 193–200.

4 I. Horrocks, A. Nechav, J. Stephenson and S. Zuberi, 'Anoxic-epileptic seizures: Observational study of epileptic seizures induced by syncopes', *Archives of Disease in Children*, 2005, vol. 90, no. 12, pp. 1283–1287.

5 J. Varasdi, *Myth Information*, p. 122.

6 S. Juan, 'Is long life related to where you live?', *The Register*, 6 June 2006.

7 S. Juan, 'It would be a lie to say that it's a lie-detector', *National Post*, 6 November 2006, pp. 1–2.

8 Dr Steven Stack is a psychologist at Wayne State University in Detroit.

9 S. Stack, 'Opera subculture and suicide for honor', *Death Studies*, 2002, vol. 26, no. 6, pp. 431–437.

10 J. Holoubek and E. Holoubek, 'Execution by crucifixion. History, methods and cause of death', *Journal of Medicine*, 1995, vol. 26, nos. 1–2, pp. 1–16.

11 Drs F.P. Retief and L. Cilliers are from the University of the Free State in Bloemfontein, South Africa.

12 F. Retief and L. Cilliers, 'The history and pathology of crucifixion', *South African Medical Journal*, 2003, vol. 93, no. 12, pp. 938–941.

13 Dr Raymond Fish is from the University of Illinois in Champaign and Dr Leslie Geddes is from Purdue University in West Lafayette, Indiana.

14 R. Fish and L. Geddes, 'Effects of stun guns and tasers', *The Lancet*, 2001, vol. 358 (9283), pp. 687–688.

15 Dr William P. Bozeman is from the Department of Emergency Medicine at the School of Medicine at Wake Forest University in Winston-Salem, North Carolina.

16 W. Bozeman, 'Withdrawal of taser electroshock devices: Too many, too soon', *Annals of Emergency Medicine*, 2005, vol. 46, no. 3, pp. 300–301.

17 J. Strote and H. Range Hutson, 'Taser use in restraint-related deaths', *Prehospital Emergency Care*, 2006, vol. 10, no. 4, pp. 447–450.

18 D. Lakkireddy, D. Wallick, K. Ryschon, M. Chung, J. Butany, D. Martin, W. Saliba, W. Kowalewski, A. Natalie and P. Tchou, 'Effects of cocaine intoxication on the threshold for stun gun induction of ventricular fibrillation', *Journal of the American College of Cardiology*, 2006, vol. 48, no. 4, pp. 805–811.

19 E. Jenkinson, C. Neeson and A. Bleetman, 'The relative risks of police use-of-force options: Evaluating the potential for deployment of electronic weaponry', *Journal of Clinical Forensic Medicine*, 2006, vol. 13, no. 5, pp. 229–241.

20 S. Juan, 'Can stun guns and tasers cause death?', *The Register*, 13 October 2006.

21 Drs S. Izumi, A. Suyama and K. Koyama are from the Radiation Effects Research Foundation in Hiroshima.

22 S. Izumi et al., 'Radiation-related mortality among offspring of atomic bomb survivors: A half-century of follow-up', *International Journal of Cancer*, 2003, vol. 107, no. 2, pp. 292–297.

23 Drs C. Hansen and C. Schriner are from the Department of Veterans Affairs in St Louis.

24 D. Hansen and C. Schriner, 'Unanswered questions: The legacy of atomic veterans', *Health Physics*, 2005, vol. 89, no. 2, pp. 153–163.

25 S. Juan, 'Survivors' kids have normal lifespan', *National Post*, 8 May 2006, p. 1.

26 Dr Troy Case is from the Department of Anthropology at North Carolina State University in Raleigh.

27 J. Varasdi, *Myth Information*, pp. 212–213.

28 T. Case, 'An analysis of scalping cases and treatment of the victims' corpses in prehistoric North America', unpublished paper, 2006.

29 S. Juan, 'What was wrong with the "bubble boy"?', *The Register*, 15 May 2006.

30 K. Creed, 'Doctors study mysterious "Burping Corpse" in LA', *Weekly World News*, 12 May 1998, p. 11.

31 S. Juan, 'Can a corpse burp?', *The Register*, 30 June 2006.

32 Dr David Lester is a professor of psychology at the Richard Stockton College in Pomona, New Jersey.

33 D. Lester, 'Personal violence (suicide and homicide), traffic fatalities and alcohol consumption', *Psychological Reports*, 1988, vol. 62, no. 2, p. 433.

34 Drs T.J. Scanlon, R. Luben, F. Scanlon and N. Singleton are from the Department of Public Health at the Mid Downs Health Authority in West Sussex, UK.

35 T. Scanlon et al., 'Is Friday the 13th bad for your health?', *British Medical Journal*, 1993, vol. 307 (6919), pp. 1584–1586.

36 Dr Simo Nayha is from the Department of Public Health Science and General Practice at the University of Oulu in Finland.

37   S. Nayha, 'Traffic deaths and superstition on Friday the 13th', *American Journal of Psychiatry*, 2002, vol. 159, no. 12, pp. 2110–2111.

38   Dr Donald Smith is a psychiatrist from Risskov, Denmark.

39   D. Smith, 'Traffic accidents and Friday the 13th', *American Journal of Psychiatry*, 2004, vol. 161, no. 11, p. 2140.

40   Drs I. Radun and H. Summala are from the Traffic Research Unit of the Department of Psychology at the University of Helsinki, Finland.

41   I. Radun and H. Summala, 'Females do not have more injury road accidents on Friday the 13th', *BMC Public Health*, 2004, vol. 4, p. 54.

42   Drs V.V. Kumar, N.V. Kumar and G. Isaacson are from the Department of Otolaryngology at the School of Medicine at Temple University in Philadelphia.

43   V. Kumar et al., 'Superstition and post-tonsillectomy hemorrhage', *Laryngoscope*, 2004, vol. 114, no. 11, pp. 2031–2033.

44   S. Juan, 'Is Friday the 13th bad for your health?', *The Register*, 7 July 2006.

45   S. Juan, 'No ill fortune on Friday the 13th', *National Post*, 16 October 2006, pp. 1–3.

46   Dr K. Eto is from the Japanese Ministry of the Environment and the National Institute for Minamata Disease in Minamata.

47   K. Eto, 'Minamata disease: A neuropathological viewpoint', *Seishin Shinkeigaku Zasshi*, 2006, vol. 108, no. 1, pp. 10–23.

48   S. Juan, 'The Minamata disaster — 50 years on', *The Register*, 14 July 2006.

49   Personal communication, 6 July 2006.

50   S. Juan, 'Can you die from testing a 9V battery on your tongue?', *The Register*, 28 July 2006.

51   Centres for Disease Control and Prevention, *Diphtheria*, Atlanta, Georgia, 2 August 2006.

52   S. Juan, 'What is diphtheria?', *The Register*, 25 August 2006.

53   S. Juan, 'What is the difference between a virus and a bacterium?', *The Register*, 29 September 2006.

54   L. Kund van der Post, 'Dive, dive, dive', *New Scientist*, 5 October 1996, p. 65.

55   M. Gregorie, R. Clifton, M. Morton, K. Bastien and A. Bowyer, 'Free-falling', *New Scientist*, 29 July 2006, p. 65.

56   S. Juan, 'From what height can you survive a dive into water?', *The Register*, 20 October 2006.

57   Dr T. Bernstein is from the Wallace-Kettering Neuroscience Institute at Wright State University in Dayton, Ohio.

58   L. Zynda and K. Skiba, 'Fracture of both humeral bones after electrocution', *Chirurgia Narzadow Ruchu I Ortopedia Polska*, 1991, vol. 56, nos. 1–3, pp. 64–65.

59   N. Friswell, M. Follows and M. Brown, 'When people die of electric shocks, what kills them — current or voltage?', *New Scientist*, 22 April 2006, p. 65.

60   S. Juan, 'What happens when you are executed by electrocution?', *The Register*, 20 October 2006.

61   Office of the Coordinator for Counterterrorism, *Country Reports on Terrorism*, US Department of State, Washington, DC, 28 April 2006.

62   Dr H.A. Sampson is from the Elliot and Roslyn Jaffe Food Allergy Institute at the Mount Sinai School of Medicine in New York.

63 H. Sampson, 'Anaphylaxis and emergency treatment', *Pediatrics*, 2003, vol. 111, no. 6, pt. 3, pp. 1601–1608.

64 World Health Organization (WHO), *The World Health Report 2002 – Reducing Risks, Promoting Healthy Life*, WHO, Geneva, 2002, p. 82.

65 S. Juan, 'What if you are hit by lightning', *National Post*, 8 May 2007, pp. 1–2.

Acknowledgements

There are so many people besides myself who have made this book possible. *Can Kissing make you Live Longer?* marks an association with HarperCollins Australia which began in 1993. Such an association is rare these days. It has been a pleasure all the way! Many thanks to Mel Cain (Publisher), Michael Moynahan CEO, Anabel Blay (International Rights), Melanie Peake (who maintains my website at HarperCollins Australia), and everyone in production, publicity and sales — many of whom I have never met but wish I could.

Thanks to Sandra Loy for editing the manuscript and doing such a great job. Thanks to the researchers, authors and publishers for the findings mentioned in this book. Thanks also to the staff at the University of Sydney libraries, the National Library of Australia in Canberra, the US Library of Congress, the US National Library of Medicine and the US National Institutes of Health in Washington, DC. Thanks to radio and TV station hosts and producers who continually invite me on to their programs to do what is so enjoyable doing.

Of course, never forgotten, are the readers who sent in questions that were used in this book: Jeanne A., Tanjina Ahmad, Gail Akami, Lucy Altmann, Ken Alvarez, Ian Anderson, Tanya Applegate, Sonia Axtens, Jackie Bainbridge, Amy Batisse, Paulo Belluomini, Rene Bernard, Judith Berry, Nikki Bertrand, Lisa Blumfield, Jon Bovard, Nikki Boyle, Anka Brautigam, Alec Burchfield, S. Burgess, Lisa Burnham, Terry Cardner, Ann Carter, Ann Cesare, Sarah Charles, Amy Charlesworth, Pat Cole, Jim Cranston, Jack Crompton, Angela Cross, Lawrence Cuthbert, Lynn Davis, Vikki de Melendez, Trina

Douglass, Liz Downes, Charlene Dupree, Michel Durinx, John Edwards, Tony Ferenczi, Elektra Filipo, Casey Filocamo, Peter Fletcher, Leanne Foster, Jenny Fredericks, Michael Friesen, Nicole Frieze, Thomas Glass, George Gomez, Hailey Goode, Amy Grambs, Ian Grant, Matt Gray, Neil Greenwood, Jeff Grisham, Nicole Haack, Gina Hadley, Rita Hamblyn, Alan Harper, Ian Hawthorne, Charles Haywood, Ville Herva, Mick Higgins, Jo Hopkins, Nikki Hunt, Luigi Imperiale, Kumari Issar, Colin Jackson, Nikki Jackson, Joel Jamal, Melanie James, Ron James, Humphrey Johnson, Jeff Johnson, Liam Johnson, S. Johnson, B. Jones, Lance Jones, Shawn Joseph, Angela Karam, Christine Koch, Jang Dae Kim, Paul King, Alison Klein, Troy Landis, Karen Landon, Andrew Lane, Nicole Lanier, Tom Lanier, Ronnie Lathrop, Ian Leonard, Chantal Liebert, Patricia Lowe, Jonathan Marten, Jill Maynard, Shauna McInerney, Lisa McMillan, Taylor Medved, Sarah Murdoch, Peter Neaum, Len Newberry, Janet Noonan, Heather Norman, Katy O'Brien, Jack O'Connor, Kelly O'Connor, Lotti Otunnu, Miguel Padilla, Jill Pascoe, Mike Pell, Andreas Pergantis, Nick Pettefair, Richard Pierce, Raoul Pimental, Lou Portman, Kelli Pritchard, Susan Quarry, John Rae, Alicia Rauzok, Kelly Reed, Kevin Reinelt, Anna Ro, Jill Rogers, Bianca Rossi, Giulia Rossi, Vincent Rot, Gerry Ryan, Kevin Ryan, Nikki Ryan, Peter Ryan, Leo Sanualio, Marc Savage, Thomas Schantz, Chuck Schroeter, Sean Scully, Tom Sherwood, Carl Smiles, Belinda Smith, Ingrid Smith, Lee Staniforth, Russell Stapleton, Karl Stefanovic, Cade Stevens, Jackie Stokes, Gerhard Strasse, Roger Sutton, Hannah Swain, Miyoki Takahashi, Ron Talbot, Charli Tricase, Ralph Turner, Mike Valentine, Felix Verity, Jason Waghorn, Jo Walker, Andrew Wallace, Leanne Ward, Karen Watterson, David Webb, Erin West, David White, Ginger Whitlam, Vanessa Wilkins, Lisa Wilkinson, Maurice Wills, Xander Winson, Andrew Wiseman, Michael Woodhams, Tommy Wooltorton, John Wright, Pat Wright, and Rodney York.

Finally, thanks to the many readers of my books and others who have sent correspondence asking questions previously answered. Especially noteworthy are the emails from young readers.

Curiosity is a great thing!

Dr Stephen Juan

About the Author

Dr Stephen Juan is the author of 'The Odd Books' series, including *The Odd Body* (1995), *The Odd Brain* (1998), *The Odd Body 2* (2000), *The Odd Sex* (2001) and *The Odd Body 3* (2007). This is Dr Stephen Juan's twelfth book.

Scientist, educator and journalist, Dr Juan is an anthropologist by training and one of the world's best communicators of research. His various books have been translated into 27 languages. He has served as a magazine editor and a newspaper columnist for the *Sydney Morning Herald*, the *Sun-Herald* (Sydney), *The National Post* (Toronto), *The New York Daily News* and *The Register* (London). A lively and popular speaker, Dr Juan appears regularly on news and current affairs television and radio programs covering any and all topics having to do with being a human being. For more than 30 years, Dr Juan taught in the Faculty of Education and Social Work at the University of Sydney. He retired in 2009, but continues as the Ashley Montagu Fellow for the Public Understanding of Human Sciences.

Index

A

Abkhasians of Georgia 241
abortions, spontaneous 13
absence 238
Acapulco, cliff divers of 257–258
accessory 120
accidents
    alcohol and 202–203
    nuclear 41–42, 247
    proneness to 234–236
acne 109–110
Adams, Scott 227
addiction 32
    gambling 228–229
    Internet Addiction Disorder (IAD) 227–228
Adherents 89
adolescent honesty 217
advertising, subliminal 213–214
Africa
    AIDS epidemic 163–164
    buttock size 182
    women carrying baskets on heads 33
agreeableness 235
AIDS *see* HIV/AIDS
air temperature 116, 192, 199–200
albinism 121–123
Albinus, Bernhard 108–109
alcohol
    accidents and 202–203
    alcohol-induced headache 36
    drinking to keep warm 201–202
    intoxication through power of
        suggestion 229–230
    metabolism of 36, 186
    thinning the blood 202
    thirst and 202

Allen, Woody 219
allicin 67–68, 69
alopecia areata 131
Alpha Centauri 44
amblyopia 48
amniotic fluid 16–17, 185
amoebic dysentery 183
anal region 103
anaphylaxis 260
ancestors 10
ancient Assyrians 242–243
ancient Egyptians 107, 153, 157
ancient Greece 138
Andaman Islanders 182
Andromeda Galaxy 44
anhedonia 218–219
Animalia 3
animals *see also* mosquito
    albinism 123
    cow's milk 62
    facial attraction 32
    handedness 142
    heart rate 157
    jumping 143
    primate 4, 132, 137, 139, 180
    prosimians 4
    sense of smell 71, 116
    shark scales and human teeth 147
    survival without head 29, 32
    sweating 107
    terminal velocity 258
    Vomeronasal Organ (VNO) 67
Anne Boleyn 126
Annie Hall (film) 219
anorexia nervosa 77
anosmia 58
anoxaemia 243

anthrax 119
anti-dandruff shampoos 42–43
antibiotics 119
antiemetics 178
apocrine glands 106
apocrine miliaria 106
appendectomy 180
Apt, Jay 45
Arabs 89
Argentina's Dirty War 232–234
argyria 119
Aristotle 110
armpits 103
Armstrong, Neil 45
arterial plaque 153–155
artificial brains 23
artificial eye 49
astronauts 189–191
atheromas 154
atherosclerosis 154–155
atomic bomb survivors 246
atomic veterans 246–247
attachment (of babies) 16
attraction
    facial 31–32
    to mosquitoes 114–117
    sexual 66–67, 99
Australia 125, 256
autonomic nervous system (ANS) 113–
    114

B
babies *see also* foetus
    amniotic fluid 16–17, 185
    birth memory 17–18
    blink rate 45–46
    breast milk 62, 171, 185
    crying *see* crying babies
    happiness of 15–16
    runny nose in 62–63
    sinuses 63
    skin 104–105
    smell of nappies 61–62
    sucking reflex 140
    tetanus 37–38
backpacks 33

bacteria
    described 256–257
    in the gut 171–172, 183
    inside the body 183–184
    on skin 103–104
bacterial meningitis 85
BadVibes project 81
balance, muscles used for 198
bathing 124–125
batteries 253–254
beards 128, 133, 224
Beethoven, Ludwig von 74–75
bends, the 189–190
Bengali speakers 88
birth labours 7
birth memory 17–18
birth order 155–156
birth sex ratio 21–22
Black Dahlia 212
bladder control 172
Blair, Tony 211
bleeding 158 *see also* menstruation
bleeding nose 64
blindness
    caused by camera flash 51–52
    inattentional 50–51
    seeing colour by touch 49–50
blinking 45–46, 56
blood
    blood alcohol concentration (BAC)
        203
    composition 158, 163
    donation 161–163
    iron in 189
    types of 117
blood banks 163
blood cells 163
blood vessels 200
Bloom, Orlando 236
body
    body denying families 141
    decomposition 250
    density 144
    life inside 183–184
    light generation by 187
    light inside 188

organs 188
in space 189–190
temperature *see* body temperature
terminal velocity 257–258
water loss 186
body cells *see* cells
body lice 103
body temperature 168–169, 192, 199–201
bone cells 24
bones
in skull 28
strength 149
teeth and 147
book binding with human skin 108–109
bottle-fed babies 62
bottoms 180–182
botulinum toxin 106
bovine gallstones 175
bowel control 172
brain *see also* memory
amoeba in 184
artificial 23
blow to nose and 63–64
cry response 15
crying babies and brain damage 14–15
dopamine release in 229
effect of alcohol on 203
feeling pain 34
smell and sight stimuli 59–60
brain cells 24
brain freeze 35
Brazil 7
bread crusts 128–129
breaking wind 173
breasts
growth 105
shrinking 196–197
size 143
breast milk 62, 171, 185
breast milk lines 120
breathing 17
garlic breath 67–68
vapourisation 165
Bristol Royal Infirmary 108–109
Brott, Boris 17–18
Brown, Harold P. 258

brushing hair 131–132
Bubble Boy (film) 248–249
Buddhism 89, 90
Burma 53
burping corpse 250
buttocks 180–182
butyric acid 176–177

C
Caesar, Julius 6–7
Caesarean section 6–8
caffeinated coffee 230
caisson sickness 189–190
callous 126
callum 126
camera flash 51–52
Campbell de Morgan spots 123–124
cancer cells 187
cancer patients 231
carbon dioxide (CO_2) 114
cardamom 68–69
cardiac cycle 155
cardiac muscle 150
Carotid Artery Disease (CARAD) 155
Carreras, José 168
Carroll, Lewis 253
carrying loads 33
cells
age 24
cancer 187
numbers 25
reproduction 242
size 26
types 24–25
type names 25
centenarians 241
Charles VII, King of France 77
chemical plants 252–253
Cheney, Dick 118
Chernobyl nuclear accident 41–42, 247
cherry angiomas 123–124
chewing gum 102
childbirth *see* birth
children
Moken children 53
mutilation of 8–9

physical development of 136–137
  sense of hearing 186
China 135, 241
Chinese restaurant headache 35
Chisov, Lt I.M. 257
cholesterol 153
cholesterol stones 174
Chordata 3
Christianity 89
chromhidrosis 106
chromosomes
  abnormal 12–14
  genes and 12
Chronic Limb Ischaemia (CLI) 154–155
chronobiology 139–140
cigarette smoking *see* smoking
circadian rhythm 139–140
circulatory system 163
circumcision 8
cirrhosis of the liver 133
clapping hands 137–138
claques 138
clavi 126
cleanliness 124–125
climate change 231–232
cloning 11–12
Clostridium tetani 37
cloth 117
coagulation 158
coffee 230
cognition 26–27
cold, feeling 199–200
colds
  age-related 62–63
  catching from wet hair 192
  feeding 169–170
  garlic cure 69
coliforms 171
Collins, Michael 45
colloidal silver 119
colour
  colour blindness 56
  effect of on smell 60
  of pubic hair 129–130
  seeing by touch 49–50
  in sweat 105–106

complex motor tic 29
complex phonic tic 29
computer power 26
computerese 88
Confessing Sam 211–213
confirmation bias 207–208
congeners 36
conspiracy theories 207–208, 211
Constantine, Emperor 243
constipation 173
consumption 164–165
Cook, Thomas L. 235–236
coprolalia 29
corn 126
corneal dryness 46
corneal erosion 55
Cornil, Dr Victor 108–109
Coronary Artery Disease (CORAD)
  155
corpse burping 250
corpuscle 158
Cri du Chat syndrome 13
Cro-Magnon humans 22
cross-eye 47–48
crucifixion 242–243
crying babies
  brain damage and 14–15
  cry response 15
  hearing damage from 75
cryptomnesia 217
culled cohort theory 22
cuticle 134
cyclical asymmetry 79–80
cyclotropia 48

D
Da Vinci Code (film) 55
damaged cohort theory 22
dandruff 42–43
danger triangle of the face 64
Darwin, Charles 73, 110–111,
  219–220
De Vita, Ted 249
deafness 74–75 *see also* hearing
decaffeinated coffee 230
decompression sickness 189–190

deep vein thrombosis (DVT) 160–161
dehydration 203
delayed alcohol-induced headache 36
deletion, chromosome 13
delusions 206
dengue fever 116
denial 230–234
dental floss 149
dental plaque 148, 183
dentin 147
deodorants 107
dependability 235
dermis 105
Descent of Man (Darwin) 73, 219–220
diabetes and garlic 70
Diana, Princess 207
diaphragm 167
diastole 155
dieting 175
digestive system 179, 193
digestive tract 183–184
Dilbert Principle, The 226–227
dimples 126
diphtheria 255–256
disgust 62
diving 257–258
DNA (deoxyribonucleic acid) 11, 12, 25,
    26 *see also* genes
doodling 208–211
dopamine 229
dorsal nerve chord 3–4
Down's syndrome 13
drinking
    to keep warm 201–203
    to stay cool 168–169
    swallowing 17, 96–97
drugs 26–27
DTP triple vaccine 255
duplication of chromosomes 13

E
E. coli 171–172
ears *see also* hearing
    cyclical asymmetry in 79–80
    earwax 80
    length in ageing 72–73

eating
    anaphylaxis 260
    bread crusts 128–129
    diet and mosquito bites 115, 116
    food allergy and kissing 86
    food poisoning 172
    morning sickness 19
    mud 102
    salmonella 179–180
    spinach 170–171
    before swimming 193
    tripe digestion 179
    worms 217–218
echolalia 29
echopraxia 29
economic values 222–223
economy class syndrome 160–161
Edison, Thomas 258
Edward's syndrome 13
electric chair 258–259
electric shock 261–262
Emerson, Ralph Waldo 211
emetology 178
emotions 153
employment prospects 223–224
endostyle 4
English speakers 88
English vocabulary 88
enhancement, human 26–27
environment and natural selection 20–21
epidermis 105
epilepsy 77
Erasmus 211
erections 187
ergot 76
erysipelas 109–110
Escherichia coli 171–172
esotropia 48
ESP (extrasensory perception) 204–205,
    214
ethanol 36
ethmoid bone 70–71
etiology 124
Eukaryota 3
eumelanin 129–130
evolution, human 5, 59

execution methods 258
exercise *see* physical activity
exocrine 194
Expression of the Emotions in Man and
   Animals, The (Darwin) 110–111
extrasensory perception (ESP) 204–205,
   214
eyes *see also* sight
   artificial 49
   blinking 45–46, 56
   cross-eye 47–48
   eye muscles 48
   eye pupil dilation 30, 56
   eye twitching 30
   eyeball orbit 56
   eyeball size 56
   eyeball socket 56
   eyelashes 186
   lazy eye 48
   ocular albinism 121–123
   Popeye Effect 170–171
   rapid eye movement (REM) sleep
      54–55
   squint eye 47–48
   walleye 47–48
   wandering eye 47–48

F
face
   cyclical asymmetry in 79–80
   danger triangle of 64
   facial attraction 31–32
   facial hair 128, 132 *see also* beards
   reading faces 132
   Whistling Face Syndrome 93
false strabismus 48
fantasy 239
farting 173
fasciculations 30
feeding a cold 169–170
feeling cold 199–200
feeling pain 34
feeling pleasure 218–219
feeling suffering 217–218
feet 107, 121, 149
   big toe 139

foot odour 143
   gender differences 142–143
   heel bone 139
   toe-tapping 138
   toenails 133–135, 184
fermentation 171–172
fertile face attraction theory 31–32
fidgeting 198
fight or flight response 188
fillers 91
fingers *see also* hands
   finger agnosia 142
   finger sacrifice rituals 8–9
   fingernails 133–135, 184, 187
   fingerprints 184
fistula 201
flight or fight response 172
flossing teeth 148
foetus *see also* babies
   body hair 127–128
   brain 185
   foetal dose effect 179
   heart rate 157
   hiccups 98
   memory 18
   numbers 7
   urethra 194
follicle mite 104
food *see* eating
Fornari, Franco 220–221
Foster, Dr George 222–223
Fox-Fordyce disease 106
free edge of nail 133
free falling 257–258
Freud, Anna 231
Freud, Sigmund 56, 209, 220, 231,
   236
Friday the 13th 250–251
frowning 126

G
Galestor, Henry 250
gallbladder 173
gallstones 173–175
galvanic skin response (GSR) 215
gambling 228–229

garlic
  as cure for diabetes 70
  as folk medicine 69
  garlic breath 67–68
  skin burn from 70
Gates, Bill 211
gender
  ability to withstand extremes 191
  anatomical differences 142, 144,
    181
  birth ratios 21–22
  death of partner 242
  doodling patterns 210
  gallstones 174–175
  heart rate 156, 157
  mosquito bite theories 114–115
  musical preferences 78–79
  urethra development 194
genes 12 see also DNA (deoxyribonucleic
    acid)
genetic-based disease 9
genetic mutation 20–21
Genghis Khan 10
geographic differences
  genetic-based disease 9
  myopia 47
  types of sweat glands 106
Germany
  birth records 21–22
  German speakers 88
  reaction to holocaust 232
gigantomastia 105
glabrous skin 113
global warming 231–232
gluteal muscles 181, 182
Goeckel, Rudlof 204
goitres 41
good, limits to 222–223
Grahl, Alex 122
Great Fire of London 212
Great Wall of China 44–45
gum disease 148
gum renewal 186
gut 171–172, 183
gut cells 24

H
haemophiliacs 159–160
hair
  albinism 121–123
  beards 128, 133, 224
  brushing 131–132
  catching colds from wet hair 192
  composition of 130
  curly or straight 128–129
  dyeing 122
  facial hair 128, 132
  foetal body hair 127–128
  hairy chests 133
  longest 130
  pubic hair colour 129–130
  turning white 130–131
hallucinations 206
hallux 139
halos 187
hands see also fingers
  hand clapping 137–138
  handedness 141–142
  photon output from 187
  thumb sucking 102, 140–141
hangovers 36
hardening of arteries 154–155
Hart, John 212
hat makers 253
head
  carrying loads on 33
  head lice 103
  headaches 34–37, 58
headphones 75
health and nails 134–135
hearing see also ears
  bad sounds 81
  deafness 74–75
  foetus memory 18
  frequencies 186
  selective mutism 99–100
  sound of voice 99
  speech and 98–99
  voices 75–78
heart
  anatomy of 151
  effect of exercise on 152–153

functioning of 156–157
location of 151
rate 155–156
shape of 150–151
size of 152
heart attack 148
heartbeat 150, 152, 157
heat capacity 117
heel bone 139
height 143
heloma 126
heroin 186
hiccups 97–98
Hindi speakers 88
Hinduism 89
Hippocrates 126
hips
    pendulum-like motion 33
    sway 181
Hiroshima 246–247
histology 147
HIV/AIDS 26, 116, 159–160
hobbies 235
Hofstadter, Douglas 226
Hofstadter's Law 226
homeostasis 169
Hominidae 4
hominoid 4
Homo erectus 4
Homo floresiensis 4
Homo habilis 5
Homo neanderthalis 4, 11–12, 73–74
Homo sapiens 3–5, 22
    Homo sapiens humanus 5
    Homo sapiens idaltu 4–5
    Homo sapiens non-humanus 5
    Homo sapiens sapiens 5
honesty 216–217
hookworms 183–184
Horney, Karen 220
Horwood, John 108–109
Hubert, Robert 212
human bot 104
human enhancement 26–27
human evolution 5, 59
human karyotype 13

humanoid robots 22–23
Hundred Years War (1337–1453) 75, 78
Hunter, Dr John 108
Hunzas of Pakistan 241
hymen 195–196
hyperopia 47
hypertropia 48
hypomelanism 121
hyposmia 58, 61
hypotropia 48
hypovolaemic shock 243

I
ice cream headache 35
identical twins 9
image of limited good 222–223
imagination 238–239
Imagined Poverty Syndrome (IPS) 221–222
immune system 248–249
implants 27
in vitro fertilisation 27
inattentional blindness 50–51
India 69, 241
infection
    disgust and threats of 62
    feeding colds and fevers 169–170
    thymus 40
influenza 170
intelligence and rejection 224–225
Internet Addiction Disorder (IAD) 227–228
intimacy zone 238
intoxication 229–230
Inupiak of Arctic 9
inversion of chromosomes 13–14
iodine 41
Iran 141
iron levels 189
iron sources 170–171
Irwin, Jim 45
Islam 89

J
Jack the Ripper 212
Jacobson's Organ 67

James Bond 121
Japanese 88, 252–253
jaw 147–148
jealousy, romantic 219–220
Jesus Christ 187
jet lag 140
Joan of Arc 75–78
Johnson, Dr Samuel 93
Jones, Stan 118–119
Judaism 89, 90
jumping 143
Jung, Carl 209

K
Kanie, Gin 10
Keats, John 211
Kennedy, John F. 207, 211
Key, Wilson Bryan 213–214
kissing
    fifty things about 82–85
    food allergy and 86
    longevity and 87
klazomania 29
knees, back of 139–140
Korea 143
Kusnetzoff, Dr Juan Carlos 232
Kutcher, Ashton 139

L
Laforgue, René 56
Laing, R.D. 56
language 73–74
    most widely spoken 87–88
    religions and 89
    vocabulary 88
    whistling languages 92
lanugo 127–128
larynx 167
Last Supper, The (da Vinci) 55
lazy eye 48
leprosy 256
lex caesarea 7
lexilalia 29
lice 103
lie detector tests 214–215
life expectancy 246

life extension 27 *see also* longevity
Life on Man (Rosebury) 103, 183
light generation by body 187
light inside body 188
light refraction 52
lightning strikes 261–262
Lincoln, Abraham 211
Linnaeus, Carl 3
lockjaw 37–39
lollies 149
long distance relationships 238
longevity 27, 87, 141, 186, 241
    los desaparecidos 233–234
Lovell, Jim 45
LSD (lysergic diethylamide) 76
lumbar puncture 36–37
lumbar region 37
lungs 164–166, 168
lunula 133

M
Mad Hatter 253
Madame Butterfly (opera) 168
Madame Butterfly Effect 242
magnetic compounds 115
magnetisation 189
magnetite 70–71
malaria 116
Mammalia 4
Mandarin speakers 87, 91
marble 117
Marie Antoinette 130
Marks, Paul 215
masking odours 115
maverick 118–119
measles 256
meconium 127–128
melanin 105, 121, 122, 129–130
melanocytes 121, 122, 130
melanosomes 130
memory 17–18, 54
    birth memory 17–18
menstruation 190–191
mental illness 239
mercury poisoning 252–253
Merlin's Disease 196–197

Merot's Syndrome 196–197
metal 118
Mexican values 222–223
migratory species 71
milia 104
milk *see* breast milk
Mill, John Stuart 109–110
Minamata disease 252–253
minimisational denial 231
mites 104
modafinil 26–27
Moken children 53
monoamine reception in brain 54
monosodium glutamate (MSG) 35, 94, 96
Monroe, Marilyn 139
mood and smell 60–61
moon 44–45
Moore's law 26
More, Thomas 130
morning sickness 18–19, 178–179
morphemes 91
mosquito 71, 116
mosquito bites 114–117
moulting 107–108
mouth
    bacteria in 183
    roof of 101–102
    tongue 90–91, 102
    yawning 101
MSG (monosodium glutamate) 35,
    94, 96
muscles
    cardiac 150
    cells 24
    eye 48
    gluteal 181, 182
    stretching 197–198
    used for balance 198
    used in speaking 30
music
    gender and 78–79
    language and 73–74
mutation 20–21
mutilation of children 8–9
myopia 46–47, 49
mysophilia 107

N
Nagasaki 246–247
nails 133–135, 184, 187
nail folds 133
nail plate 133
Naked Ape, The (Morris) 127
Napoleon 107
Narita, Kim 10
Native Americans 247–248
natural selection 20–21
navel 17
Nazi holocaust 232
Neanderthal humans 4, 11–12, 73–74
near activity 49
neonatal urticaria 104
nerve regeneration 186
nervous facial tic 29–30
Newman, Paul 56
Newton, Sir Isaac 253
nicotine 186
9/11 207, 211, 221, 231
9V batteries 253–254
nipple development 120–121
nitrogen mustard 124
nose *see also* smell
    anatomy 63–64
    bacteria on 103
    blocked 94–95
    ethmoid bone 70–71
    nose bleeds 64
    nostrils and smell 65–66
    picking 64
    runny nose 62–63
    sneezing 65
notochord 3
nuclear accidents 41–42, 247
nuclear bomb *see* atomic bomb survivors

O
OCT scans 46
ocular albinism 121–123
odour
    foot odour 143
    masking odours 115
    odorants 96
Oedipus Complex 220

onychomycosis 135
openness 235
opera lovers 242
opera singers 167–168
optimal speed 146
optimal stride 145–146
oral habits 141
Ottman siblings 11
ovulation 184
Oxford English Dictionary 88
oxygenated blood 197

P

Packard, Vance 213
pain, feeling 34
Painful Legs and Moving Toes Syndrome 138
Palaeolithic period 8
palatal rugae 101–102
palilalia 29
paralysis agitans 225
parapsychology 204–205
parasympathetic nervous system (PNS) 113–114
paraurethral gland of Skene 194
parenting 6, 14–15, 16, 141
Parkinson, C. Northcote 226
Parkinson's disease 225
Parkinson's Law 225–227
parosmia 58
parsley 68
Patau's syndrome 13
paternity tests 6
pathogenesis 124
Peele, Dr Stanton 228
pelvic girdle 181, 182
pepper 65
Pepys, Samuel 178
percepticide 232–234
Peripheral Arterial Disease (PAD) 154–155
Peter, Dr Lawrence J. 226
Peter Principle, The 226
phaeomelanin 129–130
phantosmia 58
pharyngeal pouches 4
pharyngeal slits 4

pheromones 66–67
phgonophobia 224
phonemes 91
photons 187
phrenology 79
physical activity 152–153, 177
physical activity level 200
physical appearance 223–224
physical development 136–137
physiological tetanus 38
picking your nose 64
pinky 142
piperine 65
plagiarism 216
Plato 110
pleasure 218–219
plunging 257–258
Pogue, William 45
poisoning
    food 172
    mercury 252–253
police weapons 244–245
polydactyly 139
polygraph technique 215
polymastia 120
polythelia 120
Popeye Effect 170–171
porcelain gallbladder 175
Portuguese speakers 88
post-lumbar puncture headache 36–37
post-traumatic fragmentation 78
post-traumatic stress disorder (PTSD) 219
power of rejection 224–225
power of suggestion 229–230
presbyosmia 58
Prichard, James Cowles 205
primate 4, 132, 137, 139, 180
prophylaxis 260
prostate 194
prostrate 195
protolanguage 73
pseudomamma 120
pseudostrabismus 48
psychiatry 205
psychokinesis 205
psychology 204

psychopath 205–206
psychosis 206
pubic hair colour 129–130
pubic lice 103
pubic region 103
pulmonary embolism 160
pulmonary TB 164

Q
quadruplets 11

R
race, defined 9–10
radiation damage 246–247
Ramsey, Jon Benet 212
rapid eye movement (REM) sleep 54–55
rat mite 104
Raynaud's disease 135
reading distance 49
reading faces 132
red blood cells 24
Reflex Anoxic Seizure (RAS) 240
rejection 224–225
religions 89–90
religiosity 216
Remote Personnel Assessment (RPA) 215
revenge seeking 220–221
rib cage 168
Ride, Dr Sally 191
Rigel 44
rings, chromosome 14
robotic arms 112
robots, humanoid 22–23
Romanovs 159
Romans 7, 243
romantic jealousy 219–220
Ross River fever 116
running 145–146, 166, 181
runny nose in babies 62–63
Russian speakers 88

S
saliva 85, 95
salmon patches 104–105
salmonella 179–180
salt water density 144

sapiens 4
SARS 26
scabies mite 104
scalping 247–248
Scaramanga gene 121
schizophrenia 112
Science of Parenting (Sunderland) 14–15
scotomisation 55–56
scratching 123
screwworm 104
sea gypsies 53
seborrhoeic dermatitis 43
secretiveness 236–237
seeing *see* sight
seeking revenge 220–221
selective mutism 99–100
self-concealment 236–237
senile angiomas 123–124
sense of hearing *see* hearing
sense of smell *see* smell
sense of taste *see* taste
sense of touch *see* touch
sense of vision *see* sight
Severe Combined Immunodeficiency
    (SCID) 248–249
sexual activity 106
sexual attraction 66–67, 99
sexual symbols 210
shaking palsy 225
shank 139
Sherpa women carrying baskets on heads
    33
shivering 113–114, 200
shock and hair colour 130–131
short-sightedness 46–47, 49
shouting 81, 98–99
showering 124–125
Sickle Cell Anaemia 9
sight *see also* eyes
    distance and 44
    effect of colour on smell 60
    seeing colour by touch 49–50
    seeing underwater 52–53
    smell and sight stimuli 59–60
sightlessness *see* blindness
Simon, Gabriele 50

simple denial 231
simple motor tic 29
simple phonic tic 29
Singing Neanderthals 73–74
sinus headaches 58
sinuses in babies 63
sitting on cold surfaces 192–193
situs inversus 188
skeleton 143
skin
    artificial 24
    blue-skinned people 120
    in book binding 108–109
    burns from garlic 70
    cherry angiomas 123–124
    galvanic skin response (GSR) 215
    layers 105
    microscopic life on 103–104
    moulting 107–108
    skin cells 24, 42
    skin rash 43, 109–110
    thickness 112
    water absorption by 112–113
    weight 126
skull 28
smell *see also* nose
    of babies' nappies 61–62
    effect of colour on 60
    garlic breath 67–68
    mood and ability to 60–61
    nasal dysfunction 58
    nostrils and 65–66
    pheromones 66–67
    sense of 58, 68, 71
    sight stimuli and 59–60
    smoking and ability to 61
    of sweat 107
    taste and 94–96
    of vomit 176–177
smiling 93
Smith, Dr Richard 108–109
smoking
    ability to smell and 61
    longevity and 186
sneezing 65
snoring 58–59

sound sources 186
sound waves 186
sounds, bad 81
Soviet Union 247
space films 189
Spanish 87
speaking *see also* language
    hesitations in 91
    muscles used in 30
    selective mutism 99–100
    tag questions 93–94
    voice volume 98–99
speed, optimal 146
sperm 184
spinach 170–171
spinal tap 36–37
squint eye 47–48
St Anthony's Fire 110
stalking 32
staring 32
stationary rigidity 198
steatomeria 182
steatopygia 182
stem cells 27
stomach 179
strabismus 47–48
stretching 197–198
stride, optimal 145–146
strollerisation 137
struck by lightning 261–262
stun guns 244–245
styrofoam 117–118
subcutaneous fat 105
subliminal advertising 213–214
sucking reflex 140
suffering, choosing 217–218
suffocation 242–243
sugar 149
suicide 242
Sullivan, Roy 262
sun exposure 261
sunburn 163
supernumerary 120
Suzuki, Umetaro 41
swallowing 17, 96–97 *see also*
    drinking

sweating 201
   in colours 105–106
   sweat glands 106
sweets 149
swimming
   best swimmers 143–144
   eating before 193
swimming lanes 144
sympathetic nervous system (SNS)
      113–114, 172
syndactyly 139
systole 155

T
tag questions 93–94
tapeworms 175–176, 184
Targett, Adrian 10
taser guns 244–245
taste
   loss of in ageing 95–96
   palate 101
   smell and 94–96
   taste buds 96, 185
   taste receptor cells (TRCs) 94
   of water 96
taxonomy 3–5
Tay-Sachs disease 9
teeth 147–149
tooth enamel 147
telekinesis 205
temperature
   of air see air temperature
   of body see body temperature
   of objects 117–118
Tereshkova, Valentina 191
terminal hair 128
terminal velocity 257–258
testes 184
tetanus 37–39
tetany 39
Thailand 53
therapy 236–237
thermal comfort 199–200
thiamine 41
thigh bone 149
thirst 202, 203

thixotrophy 198
thrombin 158
thumb sucking 102, 140–141
Thunderclap Headache 34
thymus 39–40, 41
thyroid 40–41
thyroid cancer 41–42
Tibet 241
tickling 110–112
   tickle response 112
   tickle torture 111
tie wearing 49
tobacco smoking see smoking
toe-tapping 138
toenails 133–135, 184
toilet pans 173
tongue 90–91, 102
tonsillectomy 251
touch
   seeing colour by 49–50
   temperature of objects 117–118
Tourette's syndrome 30
traffic accidents 260
transference denial 231
translocation of chromosomes
      13–14
Travolta, John 249
Triangulum Galaxy 44
trichology 128, 131–132
tripe digestion 179
triplets 10
Trisomy 13
trotting 146
tuberculosis (TB) 163–165
tuberculosis encephalopathy 164
twins
   birth rates 10
   finger-sucking behaviour 141
   gestation 185
   identical 9
   mosquito bites 116
twitching, eye 30
Tyndall, John 165
Tyndall Effect 165
typing 142
tyrosinase 122

U
umami 94, 96
underwater sight 52–53
Urdu 88
urethra 194
urinating 113–114
urine samples 173
US Congressional elections (2006)
    118–119

V
values, economic 222–223
vellus 128
vending machines 260
vernix 104
Vetter, David 249
Vicary, James 213
Vilcambans of Ecuador 241
virginity 195–196
virus 256
vitamin B1 41
vocabulary 88
voice
    hearing voices 75–78
    sound 99
    vocal tract 74
    voice box 167
    volume 98–99
    yelling 81
Voigt, Deborah 168
voltages 253–254
Vomeronasal Organ (VNO) 67
vomiting 176–178
vulva 181

W
walking 49
    hips sway 181
    pendulum-like motion 33
    strollerisation 137
    walking speed 146

walleye 47–48
wandering eye 47–48
wanus 142
washing 124–125
water loss in body 186, 201
water, taste of 96
weight loss 175
Western beliefs 10
Western bounties 248
Western economic values 222–223
whistling 92–93
    Whistling Face Syndrome 93
    whistling languages 92
Willis, Thomas 204
Wolf-Hirschhorn syndrome 13
Wolff, Christian 204
World War II 39
worms
    eating 217–218
    hookworms 183–184
    screwworm 104
    tapeworms 175–176, 184
wounds 142, 158
Wu speakers 88

Y
yawning 101
yelling 81, 98–99

Z
Zener cards 214
Zodiac Killer 212

www.ingramcontent.com/pod-product-compliance
Lightning Source LLC
Chambersburg PA
CBHW022138020426
42334CB00015B/957